AF446806

The IMA Volumes in Mathematics and its Applications

Volume 69

Series Editors
Avner Friedman   Willard Miller, Jr.

Institute for Mathematics and
its Applications
IMA

The **Institute for Mathematics and its Applications** was established by a grant from the National Science Foundation to the University of Minnesota in 1982. The IMA seeks to encourage the development and study of fresh mathematical concepts and questions of concern to the other sciences by bringing together mathematicians and scientists from diverse fields in an atmosphere that will stimulate discussion and collaboration.

The IMA Volumes are intended to involve the broader scientific community in this process.

Avner Friedman, Director
Willard Miller, Jr., Associate Director

* * * * * * * * * *

IMA ANNUAL PROGRAMS

1982–1983	Statistical and Continuum Approaches to Phase Transition
1983–1984	Mathematical Models for the Economics of Decentralized Resource Allocation
1984–1985	Continuum Physics and Partial Differential Equations
1985–1986	Stochastic Differential Equations and Their Applications
1986–1987	Scientific Computation
1987–1988	Applied Combinatorics
1988–1989	Nonlinear Waves
1989–1990	Dynamical Systems and Their Applications
1990–1991	Phase Transitions and Free Boundaries
1991–1992	Applied Linear Algebra
1992–1993	Control Theory and its Applications
1993–1994	Emerging Applications of Probability
1994–1995	Waves and Scattering
1995–1996	Mathematical Methods in Material Science

IMA SUMMER PROGRAMS

1987	Robotics
1988	Signal Processing
1989	Robustness, Diagnostics, Computing and Graphics in Statistics
1990	Radar and Sonar (June 18 - June 29)
	New Directions in Time Series Analysis (July 2 - July 27)
1991	Semiconductors
1992	Environmental Studies: Mathematical, Computational, and Statistical Analysis
1993	Modeling, Mesh Generation, and Adaptive Numerical Methods for Partial Differential Equations
1994	Molecular Biology

* * * * * * * * * *

SPRINGER LECTURE NOTES FROM THE IMA:

The Mathematics and Physics of Disordered Media

> Editors: Barry Hughes and Barry Ninham
> (Lecture Notes in Math., Volume 1035, 1983)

Orienting Polymers

> Editor: J.L. Ericksen
> (Lecture Notes in Math., Volume 1063, 1984)

New Perspectives in Thermodynamics

> Editor: James Serrin
> (Springer-Verlag, 1986)

Models of Economic Dynamics

> Editor: Hugo Sonnenschein
> (Lecture Notes in Econ., Volume 264, 1986)

Adam Bojanczyk George Cybenko

Editors

Linear Algebra for Signal Processing

With 37 Illustrations

Springer-Verlag

New York Berlin Heidelberg London Paris
Tokyo Hong Kong Barcelona Budapest

Adam Bojanczyk
School of Electrical Engineering
337 E&TC
Cornell University
Ithaca, NY 14853-3801
USA

George Cybenko
Dartmouth College
Thayer School of Engineering
Hanover, NH 03755-8000
USA

Series Editors:
Avner Friedman
Willard Miller, Jr.
Institute for Mathematics and its Applications
University of Minnesota
Minneapolis, MN 55455
USA

Mathematics Subject Classifications (1991): 15-06, 15-A06-23, 65-06, 65F, 65Y05-20

Library of Congress Cataloging-in-Publication Data
Linear algebra for signal processing / Adam Bojanczyk, George Cybenko,
    editors.
        p.   cm. — (The IMA volumes in mathematics and its
    applications ; v. 69)
    Papers based on lectures presented at the IMA Workshop on Linear
Algebra for Signal Processing, held at IMA, Univ. of Minnesota,
Minneapolis, April 6-10, 1992.
    Includes bibliographical references.
    ISBN 0-387-94491-5 (acid-free paper)
    1. Signal processing—Mathematics. 2. Algebras, Linear.
I. Bojanczyk, Adam. II. Cybenko, George. III. IMA Workshop on
Linear Algebra for Signal Processing (1992 : Minneapolis, Minn.)
IV. Series.
TK5102.9.L56 1995
621.382′2′015125—dc20   95-2605

Printed on acid-free paper.

Production managed by Laura Carlson; manufacturing supervised by Jeffrey Taub.
Camera-ready copy prepared by the IMA.
Printed and bound by Braun-Brumfield, Ann Arbor, MI.
Printed in the United States of America.

9 8 7 6 5 4 3 2 1

ISBN 0-387-94491-5 Springer-Verlag New York Berlin Heidelberg

Volume 41: Chaotic Processes in the Geological Sciences
Editor: David A. Yuen

Volume 42: Partial Differential Equations with Minimal Smoothness and
Applications
Editors: B. Dahlberg, E. Fabes, R. Fefferman, D. Jerison, C. Kenig,
and J. Pipher

Volume 43: On the Evolution of Phase Boundaries
Editors: Morton E. Gurtin and Geoffrey B. McFadden

Volume 44: Twist Mappings and Their Applications
Editors: Richard McGehee and Kenneth R. Meyer

Volume 45: New Directions in Time Series Analysis, Part I
Editors: David Brillinger, Peter Caines, John Geweke,
Emanuel Parzen, Murray Rosenblatt, and Murad S. Taqqu

Volume 46: New Directions in Time Series Analysis, Part II
Editors: David Brillinger, Peter Caines, John Geweke,
Emanuel Parzen, Murray Rosenblatt, and Murad S. Taqqu

Volume 47: Degenerate Diffusions
Editors: Wei-Ming Ni, L.A. Peletier, and J.-L. Vazquez

Volume 48: Linear Algebra, Markov Chains, and Queueing Models
Editors: Carl D. Meyer and Robert J. Plemmons

Volume 49: Mathematics in Industrial Problems, Part 5
by Avner Friedman

Volume 50: Combinatorial and Graph-Theoretic Problems in Linear
Algebra
Editors: Richard A. Brualdi, Shmuel Friedland, and Victor Klee

Volume 51: Statistical Thermodynamics and Differential Geometry of
Microstructured Materials
Editors: H. Ted Davis and Johannes C.C. Nitsche

Volume 52: Shock Induced Transitions and Phase Structures in General
Media
Editors: J.E. Dunn, Roger Fosdick, and Marshall Slemrod

Volume 53: Variational and Free Boundary Problems
Editors: Avner Friedman and Joel Spruck

Volume 67: Mathematics in Industrial Problems, Part 7
    by Avner Friedman

Volume 68: Flow Control
    Editor: Max D. Gunzburger

Volume 69: Linear Algebra for Signal Processing
    Editors: Adam Bojanczyk and George Cybenko

Forthcoming Volumes:

1992 Summer Program: *Environmental Studies*

1992–1993: *Control Theory*

Control and Optimal Design of Distributed Parameter Systems

Robotics

Nonsmooth Analysis & Geometric Methods in Deterministic Optimal
Control

Adaptive Control, Filtering and Signal Processing

Discrete Event Systems, Manufacturing, Systems, and Communication
Networks

1993 Summer Program: *Modeling, Mesh Generation, and
Adaptive Numerical Methods for Partial Differential Equations*

1993-1994: *Emerging Applications of Probability*

Discrete Probability and Algorithms

Random Discrete Structures

Mathematical Population Genetics

Stochastic Networks

Stochastic Problems for Nonlinear Partial Differential Equations

Image Models (and their Speech Model Cousins)

Stochastic Models in Geosystems

Classical and Modern Branching Processes

1994 Summer Program: *Molecular Biology*

FOREWORD

This IMA Volume in Mathematics and its Applications

xi

LINEAR ALGEBRA FOR SIGNAL PROCESSING

is based on the proceedings of a workshop that was an integral part of the 1991–92 IMA program on "APPLIED LINEAR ALGEBRA."

We thank Adam Bojanczyk and George Cybenko for organizing the workshop and for editing the proceedings. We also take this opportunity to thank Air Force Office of Scientific Research, National Security Agency, and the National Science Foundation, whose financial support made the workshop possible.

Avner Friedman

Willard Miller, Jr.

PREFACE

This volume contains papers based on lectures presented at the IMA Workshop on Linear Algebra for Signal Processing held at the Institute for Mathematics and its Applications, University of Minnesota, Minneapolis, April 6–10, 1992.

Signal processing applications have burgeoned this past decade. During that same time, signal processing techniques have matured rapidly and now include tools from many areas of mathematics, computer science, physics and engineering. This trend will continue as many new signal processing applications are opening up in consumer products and communications systems.

In particular, signal processing has been making increasingly sophisticated use of linear algebra on both theoretical and algorithmic fronts. Accordingly, the purpose of this workshop was to bring signal processing engineers, computer engineers, and applied linear algebraists together for an exchange of problems, theories and techniques. Particular emphasis was given to exposing broader contexts of the signal processing problems so that the impact of algorithms and hardware could be better understood.

The workshop explored five areas by having a sequence of talks devoted to the underlying signal processing problem, the algorithmic and analytic techniques and, finally, implementation issues for each area. The five areas were:

(1) updating SVD and eigendecompositions;
(2) adaptive filtering;
(3) structured matrix problems;
(4) wavelets and multirate signal processing; and
(5) linear algebra architectures (parallel/vector and other high performance machines/designs).

The workshop was attended by a group of excellent researchers. Many innovative concepts were presented in lectures as well as during less formal discussions. An account of these is given by this collection of papers.

The support of the Institute Mathematics and its Applications, University of Minnesota, which is gratefully acknowledged, contributed to the success of the meeting.

Adam Bojanczyk and George Cybenko

CONTENTS

STRUCTURED MATRICES AND INVERSES*

P. COMON†

Abstract. A matrix (and any associated linear system) will be referred to as structured if it has a small displacement rank. It is known that the inverse of a structured matrix is structured, which allows fast inversion (or solution), and reduced storage requirements. According to two definitions of displacement structure of practical interest, it is shown here that several types of inverses are also structured, including the Moore-Penrose inverse of rank-deficient matrices.

Key words. Displacement rank, Structured matrix, Töplitz, Hankel, Inverse, Schur, Moore-Penrose, Pseudo-inverse, Deconvolution.

AMS(MOS) subject classifications. 15A03, 15A06, 15A09, 15A57, 65F20, 65F30.

1. Introduction. Close to Töplitz or close to Hankel matrices appear in various areas including signal processing and automatic control (e.g. prediction of second-order nearly stationary time series). In radar or sonar (or more generally antenna processing), Töplitz matrices are encountered when far-field sources impinge an array of regularly spaced sensors after propagating through an homogeneous medium. If 2-dimensional regular arrays are utilized, then block-Töplitz matrices can be found. Other applications include optics, image processing (when the spreading function is shift invariant), differential or integral equations under certain boundary conditions and for certain discretizations (e.g. oil prospecting), seismics, geophysics, transmission lines, and communications... In general, these applications correspond to the solution of some inverse problems. When shift invariance properties are satisfied, the linear operator to invert is Töplitz, or block-Töplitz, and it is dealt with a deconvolution problem.

However, Töplitz matrices in the strict sense are rarely encountered in the real word, because the abovementioned invariance properties are not satisfied. For instance, second-order stationarity of long time series, or homogeneity of propagation media, are idealized assumptions. In antenna array processing, the decalibration of the array is the main cause of many problems, among which the deviation from Töplitz is one of the mildest ones. For instance in sonar, decalibration occurs because of the effects of pressure, temperature, and usage, among others. Another major cause of problems is the distorsion of wave fronts impinging the array due to inhomogeneity of the medium or to local turbulences (note that improvements

* This work has been carried out while the author was visiting the Institute for Mathematics and its Applications, University of Minnesota, Minneapolis, in April 1992.

† Thomson-Sintra, BP 157, F–06903 Sophia–Antipolis, Cedex, France and I3S–CNRS, 250 Av. Elnstein, F–06560 Sophia–Antipolis. The author can be reached at na.comon@na-net.ornl.edu.

can be obtained by assuming that the celerity is random with a small variance, but this is out of the scope of the present discussion). Lastly, a simple deviation from Töplitz that has been already studied is the effect of limited length of the data. The proximity to Töplitz then depends on the way the matrix is calculated: its displacement rank ranges from 2 to 4 under ideal assumptions.

Since the set of Töplitz (or Hankel) matrices is a linear space, it is easy to compute the closest Töplitz approximate of any matrix by a simple projection. However, this operation should be avoided in general, since it would destroy other important structures (e.g. just the rank). On the other hand, finding the best approximate of a matrix by another of given rank and given displacement rank is still an open problem. It is true that some simple iterative algorithms have already been proposed in the literature for the Töplitz case, but the convergence issue has not been completely covered.

Since the early works by Schur (1917), Levinson (1947), Durbin (1960), Trench (1964), and Bareiss (1969), a lot of work has been done. In particular, Kailath and others introduced in the seventies the concept of displacement rank, which allows in some way to measure a distance to Töplitz [16]. By the way, the concept of displacement rank may be seen to have some connections with integral and differential equations [18]. An excellent survey of related works can be found in [17]. Other recent investigations are also reported in [6].

It is known that a linear system $Tx = b$ can be solved with $O(n^2)$ flops if T is a $n \times n$ Töplitz matrix. If T is just only close to Töplitz, it is useful to define a displacement rank, δ, measuring a distance to the Töplitz structure [11]. Then it has been shown that the solution requires only $O(\delta n^2)$ flops, to be compared to the $O(n^3)$ complexity necessary to solve a dense linear system of general form. More recently, superfast algorithms have been proposed to solve Töplitz systems, and their complexity ranges from $O(nlog^2n)$ to $O(\alpha nlogn)$, $\alpha < n$, [1] [2] [3].

The displacement rank of a linear system is clearly related to the complexity of its solution. It has been shown in [4] [5] that this complexity reduction also holds for the calculation of various factorizations, provided the Schur algorithm is run on the appropriate block-matrix. In this paper, the displacement rank will be defined in a slightly more general framework, such that the structure of a wider class of matrices can be taken into account. In this framework, the first step in the quest of fast algorithms is to check whether the system considered has a displacement structure, and under what displacement operator its displacement rank is the smallest. Building explicitly fast algorithms taking advantage of this structure is the next question. However, our investigations are limited in this paper to the study of the displacement rank itself, and it will not be discussed how to build the corresponding fast algorithm.

The paper is organized as follows. Definitions and basic properties are given in section 2. In section 3, the structure of inverses and products of full-rank structured matrices is analyzed. Section 5 is devoted to the study of structured rank-deficient matrices, and utilizes preliminary results derived in section 4.

2. Definitions and first properties. The structure that will be considered in this paper is exclusively the *displacement structure* [16] [11]. Roughly speaking, a structured matrix is the sum of displaced versions of a unique generating matrix of small rank. For instance, sparse matrices may not have any interesting displacement structure. Displacement operators can be defined in different ways, and two definitions will be used subsequently.

DEFINITION 2.1. *For any fixed pair of matrices (Z, N) of appropriate dimension, the displacement of a matrix A with respect to displacement operator $\nabla_{Z,N}$ is defined as*

$$(2.1) \qquad \nabla_{Z,N} A = A - ZAN.$$

DEFINITION 2.2. *For any fixed pair of matrices (Z, N) of appropriate dimension, the displacement of a matrix A with respect to displacement operator $\Delta_{Z,N}$ is defined as*

$$(2.2) \qquad \Delta_{Z,N} A = ZA - AN.$$

In the remaining, matrices Z and N will be referred to as *displacement matrices*, and the pair $\{, Z, N\}$ to as the *displacement pattern*. . Once the above definitions are assumed in the primal space, then it is convenient to use the definitions below in the dual space, denoting by $(^*)$ the transposition:

$$(2.3) \qquad \begin{aligned} \nabla_{N,Z}(A^*) &= A^* - NA^*Z, \\ \Delta_{N,Z}(A^*) &= NA^* - A^*Z. \end{aligned}$$

DEFINITION 2.3. *Matrices for which any of the four displaced matrices (2.1), (2.2), (2.3) or (2.3) has a rank bounded by a value that does not depend on the size of A will be referred to as structured. This rank will be called the displacement rank of A with respect to the displacement operator considered, and will be denoted as $\delta^\nabla_{Z,N}\{A\}$, $\delta^\nabla_{N,Z}\{A^*\}$, $\delta^\Delta_{Z,N}\{A\}$, or $\delta^\Delta_{N,Z}\{A^*\}$.*

This definition is consistent with [6]. Displacement matrices Z and N are usually very simple (typically formed only of ones and zeros). Additionally, it can be seen that the displacement operator (2.1) is easily invertible as soon as *either Z or N* is nilpotent. To see this, assume that $Z^{k+1} = 0$ and explicit the displacement $\nabla_{Z,N}$ in the sum

$$\sum_{i=0}^{k} \nabla_{Z,N}\{Z^i A N^i\}.$$

Then this expression can be seen to be nothing else than A itself. For additional details, see [17] and references therein. Note that the results shown in this paper will not require a particular form for matrices Z and N (nilpotent for instance), unless otherwise specified. Other considerations on invertibility of displacement operators are also tackled in [4]. In [20], displacement operators are defined (in a manner very similar to [4]), but displacement ranks of products or pseudo-inverses are unfortunately not obtained explicitely. Lastly, other displacement structures, including (2.2), are being investigated by G.Heinig.

EXAMPLE 2.1. Denote S the so-called lower shift matrix:

$$(2.4) \qquad S = \begin{pmatrix} 0 & & & \\ 1 & \ddots & & \\ & \ddots & \ddots & \\ & & 1 & 0 \end{pmatrix}.$$

For Hankel matrices, it is easy to check out that we have

$$(2.5) \qquad \delta_{Z,N}^{\nabla}\{H\} \leq 2, \quad \text{for } (Z,N) = (S,S),$$

$$(2.6) \qquad \delta_{Z,N}^{\Delta}\{H\} \leq 2, \quad \text{for } (Z,N) = (S,S^*),$$

whereas for Töplitz matrices, we have

$$(2.7) \qquad \delta_{Z,N}^{\nabla}\{T\} \leq 2, \quad \text{for } (Z,N) = (S,S^*),$$

$$(2.8) \qquad \delta_{Z,N}^{\Delta}\{T\} \leq 2, \quad \text{for } (Z,N) = (S,S).$$

In these particular cases, the non-zero entries of displaced matrices are indeed contained only in one row and one column. These four statements hold also true if matrices Z and N are permuted. In other words,

$$\delta_{S^*,S}^{\Delta}\{H\} \leq 2, \quad \text{and} \quad \delta_{S^*,S}^{\nabla}\{T\} \leq 2.$$

It turns out that the definitions 2.1 and 2.2 yield displacement ranks that are not independent to each other. We have indeed the following

THEOREM 2.2. *For any given matrices Z, N, A, the two inequalities below hold*

$$(2.9) \qquad \delta_{Z,N}^{\nabla}\{A\} \leq \delta_{Z^*,N}^{\Delta}\{A\} + \delta_{Z,Z^*}^{\nabla}\{I\},$$

$$(2.10) \qquad \delta_{Z,N}^{\Delta}\{A\} \leq \delta_{Z^*,N}^{\nabla}\{A\} + \delta_{Z,Z^*}^{\nabla}\{I\},$$

where I denotes the identity matrix having same dimensions as A.

Proof. $\nabla_{Z,N} A = Z(Z^* A - AN) + (I - ZZ^*)A$ shows the first inequality, and $\Delta_{Z,N} A = Z(A - Z^* AN) - (I - ZZ^*)AN$ shows the second one. $\square$

EXAMPLE 2.3. If A is a circulant Töplitz matrix, e.g.,

$$A = \begin{pmatrix} a & b & c & d \\ d & a & b & c \\ c & d & a & b \end{pmatrix},$$

then it admits a displacement rank $\delta_{Z,N}^{\nabla}\{A\} = 1$ provided the following displacement pattern is assumed: $Z = S_3$ as given by (2.4), and

$$N = \begin{pmatrix} 0 & 1 & 0 & 0 \\ 0 & 0 & 1 & 0 \\ 0 & 0 & 0 & 1 \\ 1 & 0 & 0 & 0 \end{pmatrix}.$$

In this example, we also have $\delta_{Z^*,N}^{\Delta}\{A\} = 1$, which is conform to theorem 2.2.

EXAMPLE 2.4. Let A be a $m \times n$ Töplitz matrix. Define $N = S_n$, and

$$Z = \begin{pmatrix} 0 & 1 & 0 & 0 \\ 0 & 0 & \ddots & 0 \\ 1 & 0 & 0 & 1 \\ 0 & 0 & 0 & 0 \end{pmatrix}.$$

Then it can be seen that $\delta_{Z^*,N}^{\Delta}\{A\} = 2$, and $\delta_{Z,N}^{\nabla}\{A\} = 3$. This example shows that equality can occur in theorem 2.2.

EXAMPLE 2.5. If T is Töplitz $m \times n$ and H is Hankel $m \times p$, then the block matrix $(T\ H)$ has a displacement rank equal to 3 with respect to the displacement pattern $\{Z, N\} = \{S_m, S_n^* \oplus S_p\}$.

The notation $A \oplus B$ will be subsequently used when A and B are square to denote the block-diagonal matrix having A and B as diagonal blocks.

3. Displacement of various inverses and products. There is a number of situations where the displacement rank of a matrix can be quite easily shown to be small. Since our main concern is inverses, let us start with the simplest case.

THEOREM 3.1. *Let A be an invertible square matrix. Then*

$$(3.1) \quad (i)\ \delta_{Z,N}^{\Delta}\{A\} = \delta_{N,Z}^{\Delta}\{A^{-1}\},\ and\ (ii)\ \delta_{Z,N}^{\nabla}\{A\} = \delta_{N,Z}^{\nabla}\{A^{-1}\}.$$

In other words, A and A^{-1} have the same displacement rank with respect to dual displacement patterns.

To prove the theorem, it it useful to recall the following lemma.

LEMMA 3.1. *Let f and g be two linear operators, and denote E_λ^h the eigenspace of operator h associated with the eigenvalue λ. If λ is an eigenvalue of $f \circ g$, then it is also an eigenvalue of $g \circ f$. In addition, the eigenspaces have the same dimension as soon as λ is non-zero:*

$$dim\{E_\lambda^{f \circ g}\} = dim\{E_\lambda^{g \circ f}\}.$$

Proof. Assume λ is an eigenvalue of $f \circ g$. Then for some non-zero vector x, $f \circ g(x) = \lambda x$. Composing both sides by operator g immediately shows that

$$(3.2) \qquad\qquad g \circ f(g(x)) = \lambda g(x).$$

Next there are two cases: (i) if $g(x) \neq 0$, then $g(x)$ is an eigenvector of $g \circ f$ associated with the same eigenvalue λ; (ii) if $g(x) = 0$, then $f \circ g(x) = 0$ and necessarily $\lambda = 0$. Now assume without restricting the generality of the proof that $dim\{E_\lambda^{f \circ g}\} > dim\{E_\lambda^{g \circ f}\}$. Then there exists a vector x in $E_\lambda^{f \circ g}$ such that $g(x) = 0$ (since otherwise relation (3.2) would imply that g(x) is also in $E_\lambda^{g \circ f}$). Yet composing by f yields $f \circ g(x) = 0$ and consequently $\lambda = 0$. As a conclusion, if $\lambda \neq 0$, eigenspaces must have the same dimension. $\qquad\qquad\square$

Proof. *of theorem.* We have by definition $\delta_{Z,N}^\Delta\{A\} = rank\{ZA - AN\} = rank\{Z - ANA^{-1}\}$, and $\delta_{N,Z}^\Delta\{A^{-1}\} = rank\{NA^{-1} - A^{-1}Z\} = rank\{ANA^{-1} - Z\}$. But these two matrices are opposite, and therefore have the same rank. This proves (i).

Similarly since the rank does not change by multiplication by a regular matrix, we have $\delta_{Z,N}^\nabla\{A\} = rank\{A - ZAN\} = rank\{I - ZANA^{-1}\}$. On the other hand $\delta_{N,Z}^\nabla\{A^{-1}\} = rank\{A^{-1} - NA^{-1}Z\} = rank\{I - NA^{-1}ZA\}$. Now from lemma 3.1 we know that $ker\{I - f \circ g\}$ and $ker\{I - g \circ f\}$ have the same dimension. If f and g are endomorphisms in the same space, this implies in particular that $rank\{I - f \circ g\} = rank\{I - g \circ f\}$. Now applying this result to $f = ZA$, $g = NA^{-1}$ eventually proves (ii). $\qquad\qquad\square$

The proof that an invertible matrix and its inverse have the same displacement rank has been known for a long time, and proved for symmetric matrices [17]. However, the proof for general Töplitz matrices seems to have been given only recently in [6] for a displacement of type (2.1). Our theorem is slightly more general.

COROLLARY 3.1. *For any given square matrix A, let the regularized inverse be given by $R = (A + \eta I)^{-1}$, for some number η such that $A + \eta I$ is regular. Then the displacement ranks of A and R are linked by the inequality below*

$$(3.3) \qquad\qquad \delta_{N,Z}\{R\} \leq \delta_{Z,N}\{A\} + \delta_{Z,N}\{I\},$$

this inequality holding for both displacements ∇ and Δ.

Proof. Just write $\delta^{\nabla}_{N,Z}\{R\} = \delta^{\nabla}_{Z,N}\{R^{-1}\} = \delta^{\nabla}_{Z,N}\{A + \eta I\}$, and since the rank of a sum is smaller than the sum of the ranks, we eventually obtain the theorem. In order to prove the inequality for the displacement Δ, proceed exaclty the same way. $\qquad\Box$

When close to Töplitz or close to Hankel matrices are considered, the displacement matrices Z and N are essentially either the lower shift matrix S or its transposed. In such a case, it is useful to notice that

$$(3.4) \qquad \delta^{\nabla}_{S,S^*}\{I\} = \delta^{\nabla}_{S^*,S}\{I\} = 1.$$

On the other hand for any matrix Z (and S or S^* in particular):

$$(3.5) \qquad \delta^{\Delta}_{Z,Z}\{I\} = 0.$$

For a Töplitz matrix T, we have a stronger (and obvious) result, because T and $T + \eta I$ are both Töplitz.

$$\delta^{\nabla}_{S^*,S}\{R\} = \delta^{\nabla}_{S,S^*}\{T\}.$$

COROLLARY 3.2. *Let M be the 2×2 block matrix below*

$$M = \begin{pmatrix} A & B \\ C & D \end{pmatrix},$$

where A and D are square of dimension n_1 and n_2, respectively. Assume M and A are invertible. When the last $n_2 \times n_2$ block of the matrix M^{-1} is invertible, it can be written as $\bar{A}^{-1}$, where $\bar{A}$ is the so-called Schur complement of A in M: $\bar{A} = D - CA^{-1}B$. If M has a displacement rank $\delta_{N,Z}\{M\}$ with respect to a displacement pattern $\{Z, N\} = \{Z_1 \oplus Z_2, N_1 \oplus N_2\}$, where Z_i and N_i are $n_i \times n_i$ matrices, then the displacement rank of $\bar{A}$ satisfies the inequality below for both displacements ∇ and Δ:

$$(3.6) \qquad \delta_{Z_2,N_2}\{\bar{A}\} \leq \delta_{Z,N}\{M\}.$$

Proof. Applying twice the theorem 3.1, and noting that the rank of M is always larger than the rank of any of its submatrices, yield $\delta_{Z_2,N_2}\{\bar{A}\} = \delta_{N_2,Z_2}\{\bar{A}^{-1}\} \leq \delta_{N,Z}\{M^{-1}\} = \delta_{Z,N}\{M\}$. $\qquad\Box$

This kind of property has been noticed several years ago by Chun and Kailath. See for instance [4] [6]. This corollary restates it in the appropriate framework.

THEOREM 3.2. *Let A_1 and A_2 be two full-rank matrices of size $n_1 \times n_2$ and $n_2 \times n_1$, respectively, with $n_1 \leq n_2$. Then the displacement rank of the* *matrix $A_1 A_2$ is related to the displacement ranks of A_1 and A_2 for either displacement ∇ or Δ by*

$$(3.7) \qquad \delta_{Z_1,Z_2}\{A_1 A_2\} \leq \delta_{Z_1,N_1}\{A_1\} + \delta_{N_1,N_2}\{I_{n_2}\} + \delta_{N_2,Z_2}\{A_2\},$$

Proof. To prove the theorem, form the square matrix M of size $n_1 + n_2$:

$$M = \begin{pmatrix} I & A_2 \\ A_1 & 0 \end{pmatrix},$$

consider the displacement pattern $\{N_2 \oplus Z_1, N_1 \oplus Z_2\}$ and apply corollary 3.2. Again, since the displacement pattern is block-diagonal, the displaced block matrix is formed of the displaced blocks.

In the present case, the Schur complement is precisely the product $-A_1 A_2$. This proof is identical to that already proposed in [6] for particular structured matrices. $\qquad \Box$

Note that if $N_1 = N_2$, (3.5) implies $\delta^\Delta_{N_1,N_2}\{I\} = 0$. On the other hand, if $N_1 = N_2^* = S$, then $\delta^\nabla_{N_1,N_2}\{I\} = 1$ from (3.4). For particular displacement matrices Z and N, the general bounds given by theorem 3.2 may be too loose. In particular for Töplitz or Hankel matrices, the corollary below is more accurate.

COROLLARY 3.3. *Let S be the lower shift matrix defined in (2.4), T_1 and T_2 be Töplitz matrices, and H_1 and H_2 be Hankel. Then under the conditions of theorem 3.2:*

$$(3.8) \quad \text{(a) } \delta^\Delta_{S,S}\{T_1 T_2\} \leq 4, \quad \text{(b) } \delta^\Delta_{S,S}\{H_1 H_2\} \leq 4, \quad \text{(c) } \delta^\Delta_{S,S^*}\{T_1 H_2\} \leq 4,$$

$$(3.9) \quad \text{(a) } \delta^\nabla_{S,S^*}\{T_1 T_2\} \leq 4, \quad \text{(b) } \delta^\nabla_{S,S^*}\{H_1 H_2\} \leq 4, \quad \text{(c) } \delta^\nabla_{S,S}\{T_1 H_2\} \leq 4.$$

Proof. Equations (3.8) result from a combination of example 2.1 and theorem 3.2. In fact, take $Z_i = N_i = S$ for **(a)**, $Z_1 = Z_2 = N_1^* = N_2^* = S$ for **(b)**, and $Z_1 = N_1 = N_2 = Z_2^* = S$ for **(c)**.

On the other hand, if we try to apply theorem 3.2 to prove (3.9), we find a result weaker than desired, for we obtain $\delta^\nabla \leq 5$. A more careful analysis is therefore necessary. Restart the proof of theorem 3.2: if T_1 and T_2 are full rank Töplitz, the displaced block matrix $\nabla_{S \oplus S, S^* \oplus S^*} M$ has the following form:

$$\begin{pmatrix} \nabla I & \nabla T_2 \\ \nabla T_1 & 0 \end{pmatrix} = \begin{pmatrix} \begin{array}{cc} \begin{array}{cc} \text{x} & \\ & \\ & \end{array} & \begin{array}{cc} \text{x} & \text{x} \\ \text{x} & \\ \text{x} & \end{array} \\ \begin{array}{ccc} \text{x} & \text{x} & \text{x} \\ \text{x} & & \end{array} & \\ \end{array} \end{pmatrix},$$

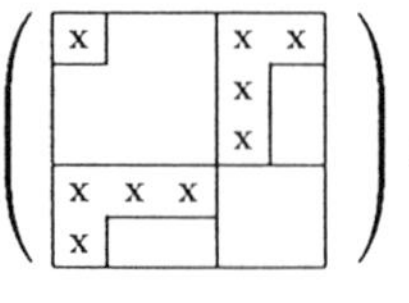

where crosses indicate the only locations where the matrix is allowed to have non-zero entries: only in two rows and two columns. Such a matrix is clearly of rank at most 4. Following the same lines as in theorem 3.2, it can be seen that the product $T_1 T_2$ has a displacement rank bounded by 4.

A similar proof could be derived in the case of two Hankel matrices, and will not be detailed here. In order to prove (3.9c), let us consider finally the block matrix

$$M = \left(\begin{array}{cc} I & H_2 \\ T_1 & 0 \end{array} \right).$$

Assuming the displacement pattern $\{S \oplus S, S^* \oplus S\}$, the displaced matrix ∇M is now of the form

$$\left(\begin{array}{cc} \nabla I & \nabla H_2 \\ \nabla T_1 & 0 \end{array} \right) = \left(\begin{array}{c} \boxed{} \end{array} \right),$$

which is again obviously of rank at most 4. The last result follows. □

The theorem 3.2 was valid only for full-rank matrices of transposed sizes. For further purposes it is useful to extend it to products of rectangular matrices of general form.

THEOREM 3.3. *Let A and B be $m \times n$ and $n \times p$ matrices. Then the product AB is also structured, and the inequality below holds:*

$$(3.10) \qquad \delta^{\Delta}_{Z_A, N_B}\{AB\} \le \delta^{\Delta}_{Z_A, N_A}\{A\} + \delta^{\Delta}_{N_A, Z_B}\{I_n\} + \delta^{\Delta}_{Z_B, N_B}\{B\},$$

where I_n denotes the $n \times n$ identity matrix.

Proof. Write first the displaced matrix as

$$\Delta_{Z_A, N_B} AB = (Z_A A - A N_A)B + A(N_A B - B N_B).$$

Then splitting the second term into $A(N_A - Z_B)B + A(Z_B B - B N_B)$ gives

$$(3.11) \quad \Delta_{Z_A, N_B}(AB) = \Delta_{Z_A, N_A} A \cdot B + A \cdot \Delta_{N_A, Z_B} I \cdot B + A \cdot \Delta_{Z_B, N_B} B,$$

which eventually proves the theorem, since the rank of a product is always smaller than the rank of each of its terms. □

A similar result holds for displacement ∇. A direct consequence of equation (3.11) is the following corollary, that looks like a differentiation rule.

COROLLARY 3.4. *When A and B have dual displacement patterns, we obtain the following simple result:*

$$\Delta_{Z,Z}(AB) = \Delta_{Z,N} A \cdot B + A \cdot \Delta_{N,Z} B.$$

In particular, if $AB = I$, this displaced matrix is null, because of (3.5).

COROLLARY 3.5. *Let A be a full-rank $m \times n$ matrix, with $m \geq n$. Then its pseudo-inverse $B = (A^*A)^{-1}A^*$ has a reduced displacement rank, as show the two bounds below:*

$$(3.12) \qquad \delta^{\Delta}_{N,Z}\{B\} \leq \delta^{\Delta}_{Z,N}\{A\} + 2\delta^{\Delta}_{N,Z}\{A^*\},$$

$$(3.13) \qquad \delta^{\Delta}_{N,Z\cdot}\{B\} \leq 3\delta^{\Delta}_{Z,N}\{A\} + \delta^{\Delta}_{Z\cdot,Z}\{I_m\}.$$

Proof. Apply corollary 3.4 to A^*A, next theorem 3.1, and lastly theorem 3.3. $\square$

EXAMPLE 3.4. If A is Töplitz, equation (3.12) claims that $\delta^{\Delta}_{S,S}\{B\} \leq 6$. In practice, it seems that no Töplitz matrix could yield a displacement rank larger than $\delta^{\Delta}_{S,S}\{B\} = 4$, which suggests that the bound is much too large.

DEFINITION 3.1. *Given any matrix A, if a matrix A^- satisfies*

$$(3.14) \qquad \begin{array}{ll} (i)\ AA^-A = A, & (ii)\ A^-AA^- = A^-, \\ (iii)\ (AA^-)^* = AA^-, & (iv)\ (A^-A)^* = A^-A, \end{array}$$

then it will be called the Moore-Penrose (MP) pseudo-inverse of A. A so-called generalized inverse need only to satisfy conditions (i) and (ii).

It is well known that A^- is unique, and that A^- and A^* have the same range and the same null space [12]. On the other hand, a generalized inverse is not unique. When a matrix A is rank deficient, it is in general not possible to construct a MP pseudo-inverse having the same displacement rank, as will be demonstrated in section 5.

4. The space of P-symmetric matrices. In this section, more specific properties shared by matrices in a wide class will be investigated. The property of P-symmetry will be necessary in section 5 to transform a matrix into its transposed just by a congruent transformation.

DEFINITION 4.1. *Let P be a fixed orthogonal n by n matrix. The set of P-symmetric matrices is defined as follows:*

$$(4.1) \qquad \mathcal{S}_P = \{M \in I\!\!R^{n \times n} / PMP^* = M^*\},$$

where $(^*)$ denotes transposition and $I\!\!R$ the set of real numbers.

It will be assumed in this section that the matrix to invert (or the system to solve) belongs to $\mathcal{S}_P$, for some given known orthogonal matrix P. For instance, if a matrix A is square and Töplitz, then it is centro-symmetric and satisfies

$$JAJ^* = A^*,$$

which shows that $A \in \mathcal{S}_J$, where J denotes the reverse identity:

$$(4.2) \qquad J = \begin{pmatrix} & & 1 \\ & \cdot^{\cdot^{\cdot}} & \\ 1 & & \end{pmatrix}.$$

If A is Hankel, then $A \in \mathcal{S}_I$ because A is symmetric. The property of P-symmetry is interesting for it is preserved under many transformations. For instance, singular vectors of a P-symmetric matrix are P-symmetric in the sense that if $\{u, v, \sigma\}$ is a singular triplet, then so is $\{Pv, Pu, \sigma\}$. A sum or a product of P-symmetric matrices is P-symmetric.

EXAMPLE 4.1. Define the 'alternate Töplitz matrix' below

$$A = \begin{pmatrix} 2 & -2 & -2 & 1 & 1 \\ 1 & -2 & -2 & 2 & 1 \\ 1 & 1 & 2 & -2 & -2 \\ 4 & -1 & 1 & -2 & -2 \\ -8 & 4 & 1 & 1 & 2 \end{pmatrix},$$

and assume the displacement pattern

$$Z = \begin{pmatrix} 0 & 0 & 0 & 0 & 0 \\ 1 & 0 & 0 & 0 & 0 \\ 0 & -1 & 0 & 0 & 0 \\ 0 & 0 & 1 & 0 & 0 \\ 0 & 0 & 0 & -1 & 0 \end{pmatrix}, \quad \text{and } N = -Z^*.$$

Then we have $PAP^* = A^*$ as requested in the definition above, with $P = J$. This matrix has displacement ranks $\delta_{Z,N}^{\nabla}\{A\} = 2$ and $\delta_{Z^*,N}^{\Delta}\{A\} = 2$, and is singular. The displacement rank of its MP pseudo inverse will be calculated in example 5.3.

PROPERTY 4.2. *The properties of P-symmetry and P^*-symmetry are equivalent.*

Proof. Let A be P-symmetric. Then transposing (4.1) gives $M = PM^*P^*$. Next pre- and post-multiplication by P^* and P, respectively, yields $P^*MP = M^*$. □

THEOREM 4.3. *If A is P-symmetric, then so is A^{-1} whenever A is invertible. If A is singular, then its Moore-Penrose inverse, A^-, is also P-symmetric.*

Proof. Inversion of both sides of the relation $PAP^* = A^*$ yields immediately $PA^{-1}P^* = A^{-1*}$. Now to insure that when A is singular, A^- is P-symmetric, it suffices to prove that the matrix $B = PA^{-*}P^*$ indeed satisfies the four conditions of definition (3.14). First, $ABA = APA^{-*}P^*A$ yields $ABA = PA^*A^{-*}A^*P^* = PA^*P = A$, which shows (i) of (3.14). Second, $BAB = PA^{-*}P^*APA^{-*}P^*$ yields similarly $BAB = PA^{-*}A^*A^{-*}P^* = $

$PA^{-*}P^*$, which equals B by definition. Next to prove (iii), consider $AB = APA^{-*}P^*$, which gives after premultiplication by PP^*: $AB = PA^*A^{-*}P^*$. But since $(A^-A)^* = A^-A$, we have $AB = PA^-AP^*$. Then insertion of P^*P yields finally $AB = PA^-P^*A^*$, which is nothing else then B^*A^*. The proof of (iv) can be derived in a similar manner. $\square$

It may be seen that in the last proof, A does not need to be a normal matrix, which was requested in a similar statement in [15]. On the other hand, it is true that if A is P-symmetric, AA^* is in general not P-symmetric.

5. Displacement of MP pseudo-inverses. In section 3, it has been shown among other things that the pseudo-inverse of a full-rank matrix is structured. It will be now analyzed how the rank deficience weakens the structure of the MP pseudo-inverse.

THEOREM 5.1. *Let A be a P-symmetric square matrix, and let Z and N be two displacement matrices linked by the relation*

$$(5.1) \qquad\qquad PZP = N.$$

Then the displacement ranks of A and A^- are related by

$$(5.2) \qquad\qquad \delta^{\nabla}_{N,Z}\{A^-\} \leq 2\delta^{\nabla}_{Z,N}\{A\}.$$

In this theorem, the condition (5.1) is satisfied in particular for both close to Töplitz and close to Hankel matrices, with $(P, Z, N) = (J, S, S^*)$ and $(P, Z, N) = (I, S, S)$, respectively.

Proof. For conciseness, denote in short δ the displacement rank $\delta^{\nabla}_{Z,N}\{A\}$, and assume A is $n \times n$. In order to prove the theorem, it is sufficient to find two full-rank $n \times n - \delta$ matrices E_1 and E_2 such that

$$(5.3) \qquad\qquad E_2^* \nabla_{N,Z}\{A^-\}E_1 = 0.$$

For this purpose, define the following full-rank matrices with n rows:

$$(5.4)\quad \begin{array}{rll} G_1 = & matrix\ whose\ columns\ span & Ker\nabla A \\ G_2 = & matrix\ whose\ columns\ span & Ker(\nabla A)^* \\ K_1 = & matrix\ whose\ columns\ span & Ker AN \cap Ker\nabla A \\ K_2 = & matrix\ whose\ columns\ span & Ker(ZA)^* \cap Ker(\nabla A)^* \\ V_1 = & matrix\ whose\ columns\ span & ANG_1 \\ V_2 = & matrix\ whose\ columns\ span & A^*Z^*G_2. \end{array}$$

Then define the two matrices E_i as:

$$E_i = [V_i,\ W_i], with\ W_i = PK_i.$$

Let us prove first that E_i are indeed of rank $n - \delta$, and then that (5.3) is satisfied.

From (5.4), we have by construction $AK_1 = 0$. Then inserting a factor P^*P and premultiplying by P gives $PAP^*PK_1 = 0$, which shows that $A^*W_1 = 0$. Yet, V_1 is in the range of A by definition, and thus V_1 and W_1 are necessarily orthogonal as members of the orthogonal subspaces $KerA^*$ and ImA. In addition, P is bijective so that W_1 has the same dimension as K_1. As a consequence, $dimE_1 = dimV_1 + dimK_1$, which is nothing else but $dimG_1$ if we look at the definitions (5.4). Similarly, one can show that W_2 and V_2 are orthogonal because $W_2 \subset KerA$. This yields after the same argumentation that $dimE_2 = dimG_2 = n - \delta$.

Now it remains to prove (5.3). To do this, it is shown that the four blocks of $E_2^*\nabla A^- E_1$ are zero. The quantity $\mu^{'} = V_2^*\nabla A^- V_1$ is null since $\mu = G_2^*ZA(A^- - NA^-Z)ANG_1$ can be written $\mu = G_2^*ZANG_1 - G_2^*ZANA^-ZANG_1$, which is the difference of two identical terms by construction of matrices G_i. In fact from (5.4), $ZANG_1 = AG_1$ and $G_2^*ZAN = G_2^*A$. Next $W_2^*\nabla A^-$ is null because $W_2^*\nabla A^*$ is null (remember that A^- and A^* have the same null space). In fact, $W_2^*\nabla A^* = K_2^*P^*A^* - K_2^*P^*NA^*Z$ by definition of W_2 and ∇. Now using the relation (5.1) and P-symmetry of A yield $W_2^*\nabla A^*P = K_2^*A - K_2^*ZAN$. These two terms are eventually null by construction of K_2. It can be proved in a similar manner that $\nabla A^*W_1 = 0$. In fact, $\nabla A^*W_1 = A^*PK_1 - NA^*ZPK_1$ implies $P^*\nabla A^*W_1 = AK_1 - P^*NP^*PA^*P^*PZPK_1$. Again these two terms can be seen to be zero utilizing (5.1), P-symmetry of A, and the definition (5.4) of K_1. $\qquad\qquad\qquad\qquad\qquad\qquad\qquad\qquad\qquad\qquad\qquad\qquad\quad\square$

This theorem is an extension of a result first proved in [7]. As pointed out in [8], when the displacement rank of A is larger than its rank, the theorem above gives too weak results as is next shown.

THEOREM 5.2. *Let A be a square matrix, and denote by $r\{A\}$ its rank. Then there exist two other bounds for the displacement rank of its MP pseudo-inverse:*

$$(5.5) \qquad \delta_{N,Z}^{\nabla}\{A^-\} < 2r\{A\} \quad if \quad \delta_{Z,N}^{\nabla}\{A\} < 2r\{A\}, \quad and$$

$$(5.6) \qquad \delta_{N,Z}^{\nabla}\{A^-\} \leq 2r\{A\} \quad otherwise.$$

Proof. The proof of (5.6) is easy. In fact, it holds true for any matrix M since

$$rank\{M - ZMN\} \leq rank\{M\} + rank\{ZMN\} \leq 2rank\{M\}$$

is always true. So let us prove (5.5). Since A has rank $r\{A\}$, it may be written as

$$A = U\Sigma V^*,$$

where Σ is invertible and of size $r\{A\}$. Define the matrices $A = [U \; ZU]$, $B = [V \; N^*V]$, and $\Lambda = Diag(\Sigma, -\Sigma)$. Then it may be seen that

$$\nabla A = A\Lambda B^*.$$

Since Λ is of full rank, either A or B must be rank defficient, otherwise ∇A would be of rank $2rank\{A\}$ which is contrary to the hypothesis. Thus assume without restricting the generality of the proof that $rank\{A\} < 2r\{A\}$. Then $rank\{\nabla A^-\} < 2r\{A\}$ because:

$$\nabla A^{-*} = A\Lambda^{-1}B^*.$$

This completes the proof. $\square$

COROLLARY 5.1. *Let T and H be close to Töplitz and close to Hankel square matrices, respectively. Then*

$$(5.7) \qquad \delta^{\nabla}_{S^*,S}\{T^-\} \le 2\delta^{\nabla}_{S,S^*}\{T\}, \ \ \delta^{\nabla}_{S,S}\{H^-\} \le 2\delta^{\nabla}_{S,S}\{H\},$$

$$(5.8) \quad \delta^{\Delta}_{S,S}\{T^-\} \le 2\delta^{\Delta}_{S,S}\{T\} + 1, \ \ \delta^{\Delta}_{S^*,S}\{H^-\} \le 2\delta^{\Delta}_{S,S^*}\{H\} + 1.$$

Proof. To prove (5.7), simply use theorem 5.1 and relation (3.4). In order to prove equations (5.8), utilize theorem 2.2 and relation (3.5). $\square$

Note that the bounds are tight enough to be reached, as now shown in examples.

EXAMPLE 5.3. Take again the matrix defined in example 4.1. This matrix is of rank 4 and displacement rank 2. In addition, the displacement pattern satisfies $PZP = N$ as required in the theorem 5.1. With the notations of example 4.1, the MP pseudo inverse of A has displacement ranks $\delta^{\Delta}_{N,Z^*}\{A^-\} = 4$ and $\delta^{\nabla}_{N,Z}\{A^-\} = 4$. This is consistent with theorem 5.1.

EXAMPLE 5.4. Define the 5×5 Töplitz matrix of rank 3 :

$$A = \begin{pmatrix} 2 & 4 & 3 & 1 & 2 \\ 1 & 2 & 4 & 3 & 1 \\ 3 & 1 & 2 & 4 & 3 \\ 4 & 3 & 1 & 2 & 4 \\ 2 & 4 & 3 & 1 & 2 \end{pmatrix},$$

and assume as displacement pattern $Z = S_5$ and $N = S_5^*$. Then A has a displacement rank $\delta^{\nabla}_{Z,N} = 2$, and its MP pseudo-inverse has a displacement rank $\delta^{\nabla}_{N,Z}$ equal to 4. This result was expected, according to corollary 5.1.

EXAMPLE 5.5. If H is Hankel, then the displacement rank of H^- with respect to the displacement operator $\Delta_{S^*,S}$ is bounded by 5.

Other particular examples can be found in [7] and [8]. Let us now switch to the case of rectangular and rank-deficient structured matrices. In order to extend corollary 3.5, we need a variant of the inversion lemma:

LEMMA 5.1. *Let M be the block matrix:*

$$M = \begin{pmatrix} P & A_2 \\ A_1 & 0 \end{pmatrix},$$

where P is square invertible, and where A_1 and A_2 have the same rank. Then the MP-pseudo-inverse of M is:

$$M^- = \begin{pmatrix} Y & -P^{-1}A_2 X \\ -X A_1 P^{-1} & X \end{pmatrix},$$

where $X = -(A_1 P^{-1} A_2)^-$, and $Y = P^{-1} + P^{-1}A_2 X A_1 P^{-1}$.

Proof. Let $A_i = U_i D_i V_i^*$ denote the SVD of A_i. Then define the matrix

$$\bar{M} = \begin{pmatrix} U_2^* & 0 \\ 0 & U_1^* \end{pmatrix} M \begin{pmatrix} V_1 & 0 \\ 0 & V_2 \end{pmatrix},$$

and apply the usual inversion lemma to the invertible square portion of $\bar{M}$, denoted B. In other words we have:

$$\bar{M} = \begin{pmatrix} B & 0 \\ 0 & 0 \end{pmatrix} \quad \text{and} \quad \bar{M}^- = \begin{pmatrix} B^{-1} & 0 \\ 0 & 0 \end{pmatrix}.$$

The last lines of the proof are then just obvious manipulations. $\square$

COROLLARY 5.2. *Let A be an $m \times n$ rectangular matrix with $m > n$. Then the displacement rank of its MP pseudo inverse verifies*

$$\delta_{N,Z}^{\Delta}\{A^-\} \le 3\delta_{N,Z}^{\Delta}\{A^*\} + 2\delta_{Z,N}^{\Delta}\{A\} + 2\delta_{N,N*}^{\nabla}\{I_m\}.$$

Proof. Write A^- as $(A^*A)^- A^*$, apply theorem 5.1 to the square matrix (A^*A), and then apply the product rule given in corollary 3.5. $\square$

6. Concluding remarks. In this paper various aspects of the displacement rank concept were addressed in a rather general framework. In particular, displacement properties of rank-deficient matrices were investigated. However the bounds given in corollaries 3.5 and 5.2 are obviously too large. It is suspected that corollary 5.2 could be improved to $\delta\{B\} \le 2\delta\{A\} + \delta\{I\}$ in most cases. On the other hand, particular examples have been found showing that the bounds given in other theorems are indeed reached (in particular theorems 5.1 and 5.2).

Another major limitation of this work lies in the fact that our proofs are in general not constructive, in the sense that they do not define suitable algorithms having the expected complexity. This is now the next question to answer. First ideas in this direction can be found in [4] and [14] and could be used for this purpose.

The author thanks Georg Heinig for his proofreading of the paper.

REFERENCES

[1] G. Ammar and W.B. Gragg, Superfast solution of real positive definite Töplitz systems, *SIAM Journal Matrix Analysis.*, vol.9, jan 1988, 61-76.

[2] A.W. Bojanczyk, R.P. Brent, and F.R. DeHoog, QR factorization of Töplitz matrices, *Numerische Mathematik,* vol.49, 1986, 81-94.

[3] R. Chan and G. Strang, Töplitz equations by conjugate gradients with circulant preconditioner, *SIAM Jour. Sc. Stat. Comput.* vol.10, jan 1989, 104-119.

[4] J. Chun, Fast array algorithms for structured matrices, *PhD thesis,* Stanford University, June 1989.

[5] J. Chun, T. Kailath, and H. Lev-Ari, Fast parallel algorithms for QR and triangular factorization, *SIAM Jour. Sci. Stat. Comput.,* vol.8, nov 1987, 899-913.

[6] J. Chun and T. Kailath, Displacement structure for Hankel- and Vandermonde-like matrices, *Signal Processing Part I: Signal Processing Theory, IMA Volumes in Mathematics and its Applications,* vol. 22, Springer Verlag, 1990 pp. 37–58.

[7] P. Comon and P. Laurent-Gengoux, Displacement rank of generalized inverses of persymmetric matrices, *Thomson Sintra report,* 90-C570-191, October 1990, to appear in *SIAM Journal Matrix Analysis.*

[8] P. Comon, Displacement rank of pseudo-inverses, *IEEE Int. Conf. ICASSP,* march 1992, San Francisco, vol.V, 49-52.

[9] P. Delsarte, Y.V. Genin, and Y.G. Kamp, A generalization of the Levinson algorithm for hermitian Töplitz matrices with any rank profile, *IEEE Trans ASSP,* vol.33, aug 1985, 964-971.

[10] K. Diepold and R. Pauli, Schur parametrization of symmetric matrices with any rank profile, *IEEE Int. Conf. ICASSP,* march 1992, San Francisco, vol.V, 269-272.

[11] B. Friedlander, M. Morf, T. Kailath, and L. Ljung, New inversion formulas for matrices classified in terms of their distance from Töplitz matrices, *Linear Algebra Appl.,* vol.27, 1979, 31-60.

[12] G.H. Golub and C.F. Van Loan, *Matrix computations,* Hopkins, 1983.

[13] C. Gueguen, An introduction to displacement ranks, *Signal processing XLV,* Lacoume, Durrani, Stora editors, Elsevier, 1987, 705-780.

[14] G. Heinig and K. Rost, Algebraic methods for Töplitz-like matrices and Operators, *Birkhäuser,* 1984.

[15] R.D. Hill, R.G. Bates, and S.R. Waters, On perhermitian matrices, *SIAM Journal Matrix Analysis,* April 1990, pp. 173–179.

[16] T. Kailath, A. Viera, and M. Morf, Inverses of Töplitz operators, innovations, and orthogonal polynomials, *SIAM Review,* 20, 1978, pp. 106–119.

[17] T. Kailath, Signal processing applications of some moment problems, *Proceedings of Symposia in Applied Mathematics, American Mathematical Society,* vol.37, 1987, pp. 71–109.

[18] T. Kailath, Remarks on the origin of the displacement-rank concept, *Applied Math. Comp.,* 45, 1991, pp. 193–206.

[19] S. Pombra, H. Lev-Ari, and T. Kailath, Levinson and Schur algorithms for Töplitz matrices with singular minors, *Int. Conf. ICASSP,* april 1988, New York, 1643-1646.

[20] D. Wood, Extending four displacement principles to solve matrix equations, submitted to *Math. Comp.,* preprint April 1992.

STRUCTURED CONDITION NUMBERS FOR LINEAR MATRIX STRUCTURES*

I. GOHBERG[†] AND I. KOLTRACHT[‡]

Abstract. Formulas for condition numbers of differentiable maps restricted to linearly structured subsets are given. These formulas are applied to some matrix maps on Toeplitz matrices. Other matrix examples are also indicated.

Key words. Linear structure, structured condition number, Toeplitz matrix.

AMS(MOS) subject classifications. 65F35, 15A12.

1. Introduction. In this paper we consider structured condition numbers for some matrix maps on linearly structured classes of matrices, notably, Toeplitz matrices, which appear frequently in signal processing, see, for example, T. Kailath [10] and references therein.

To illustrate usefulness of structured condition numbers consider the matrix inversion map at the Hilbert matrix $A = \left((i+j-1)^{-1} \right)_{i,j=1}^{10}$. It's condition number which corresponds to perturbations of A in the set of all nonsingular matrices is $3 \cdot 10^{12}$ which is also equal to the condition number at A with respect to perturbations in the set of nonsingular Hankel matrices (see Section 4 for definition). As a consequence, if one attempts to invert A on a computer with unit round-off error $u > 10^{-12}$ using a general matrix solver, or a special algorithm defined on nonsingular Hankel matrices only, then one may expect the loss of all significant figures in A^{-1}. This was indeed observed in numerical experiments in Gohberg, Kailath, Koltracht and Lancaster [2]. On the other hand, the condition number of A with respect to perturbation in the class of Cauchy matrices, (matrices of the form $\left((t_i - s_j)^{-1} \right)_{i,j=1}^{n}$), is ≤ 740. Therefore one may expect that a stable special algorithm defined on nonsingular Cauchy matrices only will give an accurate inverse of A. Supporting numerical evidence can be found in [2], and an explanation in Gohberg and Koltracht [3]. The numerical instability of a general matrix solver, or a Hankel solver, is understandable, namely, entries $(i+j-1)^{-1}$ are formed, thus introducing an ill-conditioned step in the course of solving a well conditioned problem.

We used for this illustration mixed structured condition numbers of A, (see Section 2, or Gohberg and Koltracht [4], for definition). For discussion of numerical stability of algorithms in general, we refer to Stoer and

---

* This work was partly supported by the NSF (Grant DMS-9007030)

† Department of Mathematics, Tel Aviv University, Ramat Aviv, Tel Aviv, 69978 Israel.

‡ Department of Mathematics, University of Connecticut, Storrs, Connecticut 06269-3009, USA.

Bulirsch [13] and Golub and Van Loan [8].

We remark that the Cauchy structure for which we have such a large difference between structured and general condition numbers is not linear. For linear structures we expect that there will be little difference between the two condition numbers, although we can prove it for positive definite Toeplitz matrices only, see Section 3 below. Thus general purpose stable algorithms remain (forward) stable on positive definite Toeplitz matrices.

In Section 2 we give formulas for linear structured condition numbers based on explicit representation of a linear structure and a directional derivative of a map. In Section 3 we apply these formulas to some matrix maps at Toeplitz matrices. In Section 4 we give more examples of directional derivatives of some useful matrix maps, and of some linear structures other than Toeplitz. We follow concepts and definitions of [4]. A different approach to structured perturbations of matrices can be found in Higham and Higham, [9].

2. Linear structures. Let $G : R^p \to R^q$ be a differentiable map defined on an open subset of R^p, D_G. The *usual* condition number of the map G at a point $A \in D_G, A \neq 0, G(A) \neq 0$, is given by:

$$(2.1) \qquad k(G, A) = \frac{\| G'(A) \|\| A \|}{\| G(A) \|},$$

where $\|A\|$ is some norm on $R^p, \|G(A)\|$ is some norm on R^q, and $\|G'(A)\|$ is the corresponding operator norm of $G'(A)$, as a linear map from R^p to R^q. The *mixed* condition number of G at A is defined as follows:

$$(2.2) \qquad m(G, A) = \frac{\|G'(A) D_A\|_\infty}{\| G(A) \|_\infty},$$

where $A = (A_1, \cdots, A_p)$ and $D_A = diag\{A_1, \cdots A_p\}$. The mixed condition number relates normwise errors in $G(A)$ to componentwise errors in A, hence the term: mixed. To be more specific let X_i be the perturbed value of A_i such that

$$|X_i - A_i| \leq \epsilon |A_i|, \; i = 1, \ldots, p.$$

Then

$$\frac{\|G(X) - G(A)\|_\infty}{\|G(A)\|_\infty} \leq m(G, A)\epsilon + o(\epsilon).$$

Note that zero entries of A are not perturbed, so that X preserves the sparseness pattern of A. For a more detailed discussion of the condition numbers $k(G, A)$ and $m(G, A)$ see [4]. It is clear that if $k(G, A)$ is taken with respect to the ∞- norm in R^p and R^q then

$$(2.3) \qquad\qquad m\left(G, A\right) \le k\left(G, A\right).$$

A *structured* subset of D_G is the range of another differentiable map, say, $H : R^n \to R^p$ with $n < p$. A *structured* condition number of G with respect to this structure is defined as the (usual or mixed) condition number of the restriction of G onto the structured subset, or more formally, the structured condition number of G at $A = Ha$ is the condition number of $F = G \circ H$ at a, with the notation

$$m\left(F, a\right) = \mu\left(G, A\right),$$

$$k\left(F, a\right) = \kappa\left(G, A\right).$$

In this paper we only consider the case when H is a linear map, namely, when for $a = (a_1, \ldots, a_n) \in D_H$,

$$Ha = a_1 h_1 + a_2 h_2 + \cdots + a_n h_n,$$

where $h_1, \ldots, h_n$ are some fixed vectors in R^p. We identify H with its matrix in standard bases of R^n and R^p, such that $h_1, \ldots, h_n$ are the columns of H. For example, if $H : (a_1, \ldots, a_n) \to diag\{a_1, \ldots, a_n\}$ then h_k is an $n \times n$ matrix with 1 in $(k, k) - th$ position, and zeros elsewhere, identified with a vector in R^{n^2}, (here $p = n^2$).

It follows from the chain rule of differentiation that the partial derivative of F with respect to a_k equals to the directional derivative of G with respect to h_k,

$$\frac{\partial F}{\partial a_k} = \frac{\partial G}{\partial h_k}, \ \ k = 1, \ldots, n,$$

which is a vector in R^q. Therefore $F'\left(a\right) = \left[\frac{\partial G}{\partial h_1}, \ldots, \frac{\partial G}{\partial h_n}\right]$, and hence

$$(2.4) \qquad\qquad \kappa\left(G, A\right) = \frac{\left\|\left[\frac{\partial G}{\partial h_1}, \ldots, \frac{\partial G}{\partial h_n}\right]\right\| \|a\|}{\|G\left(A\right)\|},$$

$$(2.5) \qquad\qquad \mu\left(G, A\right) = \frac{\left\|\left[a_1 \frac{\partial G}{\partial h_1}, \ldots, a_n \frac{\partial G}{\partial h_n}\right]\right\|_\infty}{\|G\left(A\right)\|_\infty}.$$

It is clear that if $\kappa(G, A)$ is taken with respect to the ∞-norm in R^n and R^q then

$$(2.6) \qquad\qquad \mu\left(G, A\right) \le \kappa\left(G, A\right).$$

Suppose now that H is an isometry. In this case it is easy to see that

$$(2.7) \qquad\qquad \kappa\left(G, A\right) \le k\left(G, A\right).$$

Indeed, since $H'\left(a\right) = H$ for any a, it follows that $F'\left(a\right) = G'\left(A\right) H$ where $Ha = A$, and hence $\|F'\left(a\right)\| \le \|G'\left(A\right)\|$. To obtain a similar inequality for mixed condition numbers we make an assumption about H which is satisfied for all linear structured classes of matrices considered in this paper.

PROPOSITION 2.1. *Suppose that the columns of $H, h_1, \ldots, h_n$, have entries equal to zero or one only. Furthermore, suppose that $h_1, \ldots, h_n$ are mutually orthogonal (or equivalently, that indices of 1's in $h_1, \ldots, h_n$ are mutually disjoint). Then*

$$(2.8) \qquad\qquad \mu\left(G, A\right) \le m\left(G, A\right).$$

Proof. We need to show that the infinity norm of $F'\left(a\right) D_a = G'\left(A\right) H D_a$ is less than that of $G'\left(A\right) D_A$. Observe that D_A is a $p \times p$ diagonal matrix whose diagonal entries are $a_1, \ldots, a_n$ in some order and with repetitions, (recall that $n < p$). Next note that $H D_a$ is a $p \times n$ matrix whose k-th column equals to the sum of all columns of D_A which contain a_k as an entry. Therefore the k-th column of $G'\left(A\right) H D_a$ equals to the sum of all those columns of $G'\left(A\right) D_A$ which have indices of those columns of D_A which contain a_k as an entry. Since each column of $F'\left(a\right) D_a$ is a sum of some columns of $G'\left(A\right) D_A$ such that each column of $G'\left(A\right) D_A$ is used exactly once, it follows that $\|F'\left(a\right) D_a\|_\infty \le \|G'\left(A\right) D_A\|_\infty$. $\square$

We see from (2.4) and (2.5) that in order to find a structured condition number of G at $A = Ha$, given the structure map H, one needs directional derivatives of G. In the next section we consider some matrix maps with known directional derivatives and find their structured condition numbers at Toeplitz matrices.

3. Symmetric Toeplitz matrices. In this section G is a map defined on $n \times n$ matrices and $H : R^n \to R^{n \times n}$,

$$H\left(a_1, \cdots, a_n\right) = A = \begin{bmatrix} a_1 & a_2 & \ddots & a_n \\ a_2 & a_1 & \ddots & \ddots \\ \vdots & \ddots & \ddots & a_2 \\ a_n & \ddots & a_2 & a_1 \end{bmatrix}.$$

We identify $R^{n \times n}$ and R^{n^2} using any fixed ordering of matrix elements, say row by row. Thus k-th column of H, h_k, is an element of R^{n^2} which corresponds to the $n \times n$ matrix with ones in positions of a_k in A and zeros elsewhere, e.g. h_1 corresponds to the identity matrix. Next we consider some specific maps defined on $R^{n \times n}$.

3.1. Matrix inversion, $G : A \to A^{-1}$. The directional derivative of G in the direction h is given by ([8], Section 2.5):

$$\frac{\partial G}{\partial h} = -A^{-1} h A^{-1},$$

and hence $F'(a) = -\left[A^{-1} h_1 A^{-1}, \cdots, A^{-1} h_n A^{-1}\right]$. Since the (i,j)-th entry of $A^{-1} h_k A^{-1}$ equals to $c_i^T h_k c_j$ where c_i is the i-th column of A^{-1}, it follows that

$$(3.1) \qquad \|F'(a)\|_\infty = \max_{i,j=1,\ldots,n} \sum_{k=1}^{n} \left| c_i^T h_k c_j \right|,$$

$$(3.2) \qquad \|F'(a) D_a\|_\infty = \max_{i,j=1,\ldots,n} \sum_{k=1}^{n} \left| a_k c_i^T h_k c_j \right|.$$

The corresponding condition numbers are now readily obtained. We remark that the computation of $\|F'(a)\|_\infty$ or $\|F'(a) D_a\|_\infty$ requires here $O\left(n^4\right)$ flops. This can be reduced to $O\left(n^3 \log n\right)$ by the use of FFT. If only one column of A^{-1} is required, e.g. the last column which gives the solution of Yule-Walker equations, then for the corresponding map, $F_n : a \to c_n$ we have

$$\|F_n'(a)\|_\infty = \max_{i=1,\ldots,n} \sum_{k=1}^{n} \left| c_i^T h_k c_n \right|.$$

This can be computed in $O\left(n^2 \log n\right)$ flops, see Gohberg, Koltracht and Xiao [6]. When A is positive definite, $\kappa(G, A)$ and $\mu(G, A)$ can be estimated faster, with the speed of solving $Ax = b$.

PROPOSITION 3.1. *Let A be a positive definite Toeplitz matrix and let G be the map of matrix inversion. Then*

$$\mu(G, A) \leq \left\{ \begin{array}{c} \kappa(G, A) \\ m(G, A) \end{array} \right\} \leq k(G, A) \leq n^2 \mu(G, A),$$

where $k(G, A) = \|A\|_\infty \left\|A^{-1}\right\|_\infty$ and $\kappa(G, A)$ is taken with respect to the infinity norm in R^n and R^{n^2}.

Proof. All inequalities, except for the last one are just (2.3), (2.6), (2.7) and (2.8). To prove the last one denote $A^{-1} = (\sigma_{ij})_{i,j=1}^{n}$ and let

$$\sigma_{mm} = \max_{i,j=1,\ldots,n} \left| \sigma_{ij} \right|.$$

Thus $\left\|A^{-1}\right\|_\infty \le n\sigma_{mm}$ and $\|A\|_\infty \le na_1$. On the other hand

$$\left\|F'(a)D_a\right\|_\infty = \max_{i,j=1,\dots,n} \sum_{k=1}^{n} \left|a_k c_i^T h_k c_j\right| \ge \sum_{k=1}^{n} \left|a_k c_m^T h_k c_m\right| \ge a_1 c_m^T c_m \ge a_1 \sigma_{mm}^2.$$

Since the norm of $A^{-1} = G(A)$ as a vector in R^{n^2} equals to σ_{mm} it follows that

$$\mu(G,A) = \frac{\left\|F'(a)D_a\right\|_\infty}{\|G(A)\|_\infty} \ge a_1 \sigma_{mm} \ge \frac{\|A\|_\infty \left\|A^{-1}\right\|_\infty}{n^2}.$$

$\square$

It can be seen from the above proof that the factor n^2 in the last inequality is a result of a sequence of rude estimates. Moreover, a large number of experiments accompanying those reported in [6] and [7] show that the ratio of $\mu(G,A)$ and $k(G,A)$ is of order unity. Also, if $A^{-1} = \left|A^{-1}\right|$ where $|\cdot|$ denotes array of absolute values, then in fact, $\mu(G,A) = m(G,A)$. Indeed, in this case, for all i and j,

$$\sum_{k=1}^{n} \left|a_k c_i^T h_k c_j\right| = \sum_{k=1}^{n} |a_k| c_i^T h_k c_j = c_i^T \left[\sum_{k=1}^{n} |a_k| h_k\right] c_j = c_i^T |A| c_j.$$

Hence $\left\|F'(a)D_a\right\|_\infty = \left\|A^{-1}|A|A^{-1}\right\|_v$ where $\|\cdot\|_v$ equals to the largest absolute value among entries of an array. Since

$$m(G,A) = \frac{\left\|A^{-1} \cdot |A| \cdot A^{-1}\right\|_v}{\left\|A^{-1}\right\|_v}$$

(see, for example [4]), it follows that $\mu(G,A) = m(G,A)$. On the basis of all this evidence we claim that for practical purposes all condition numbers of Proposition 2 are equal to each other. Thus one can estimate $\left\|A^{-1}\right\|_\infty$ instead of (3.1) or (3.2) which can be done with the speed of solving $Ax = b$, see Dongarra, Bunch, Moler and Steward [1] for a lower bound and Koltracht and Lancaster [11], for an upper bound.

The analysis of this section remains true for banded Toeplitz matrices. The only difference would be that the upper summation limit, n, in (3.1) or (3.2) is replaced by the bandwidth.

3.2. Solution of $Ax = b$. It is convenient to consider here $G = G_1 \bigoplus G_2$ defined on $R^{n \times n} \oplus R^n$, such that

$$G[A,b] = x,$$

where $Ax = b$. Instead of one condition number we suggest to use a pair, corresponding to G_1 and G_2 respectively. For example,

$$k(G,[A,b]) = [k(G_1,A), k(G_2,b)] = \left[\|A\|\left\|A^{-1}\right\|, \frac{\left\|A^{-1}\right\|\|b\|}{\left\|A^{-1}b\right\|}\right].$$

This pair has the following meaning. If $\|\tilde{A} - A\| \leq \epsilon_1 \|A\|$ and $\|\tilde{b} - b\| \leq \epsilon_2 \|b\|$ then

$$\frac{\|\tilde{x} - x\|}{\|x\|} \leq k\left(G_1, A\right) \epsilon_1 + k\left(G_2, b\right) \epsilon_2 + o\left(\max\left(\epsilon_1, \epsilon_2\right)\right).$$

By [8] Section 2.5, for any direction h in R^{n^2} we have

$$\frac{\partial G_1}{\partial h} = -A^{-1}hx.$$

Therefore $G'\left(A\right) = -\left[A^{-1}h_1 x, \ldots, A^{-1}h_n x\right]$ and

$$\|G'\left(A\right)\|_\infty = \max_{i=1,\ldots,n} \sum_{k=1}^{n} \left|c_i^T h_k x\right|,$$

$$\|G'\left(A\right) H D_a\|_\infty = \max_{i=1,\ldots,n} \sum_{k=1}^{n} \left|a_k c_i^T h_k x\right|.$$

These norms can be computed in $O\left(n^2 \log n\right)$ flops as explained in [6]. Perturbations in b are not structured and the corresponding condition numbers are

$$k\left(G_2, b\right) = \frac{\left\|A^{-1}\right\| \|b\|}{\|A^{-1}b\|},$$

$$m\left(G_2, b\right) = \frac{\left\|\left|A^{-1}\right| |b|\right\|_\infty}{\|A^{-1}b\|_\infty},$$

where $|\cdot|$ denotes array of absolute values. For $m\left(G_2, b\right)$ see Skeel [12].

The relation between $\kappa, \mu\left(G_1, A\right)$ and $k, m\left(G_1, A\right)$ requires additional study.

3.3. A Simple eigenvalue. Let $G : A \to \lambda \neq 0$, where λ is a simple eigenvalue of A. Let x be the appropriately normalized eigenvector. Then, [8] Section 7.2,

$$\frac{\partial G}{\partial h} = x^T hx,$$

and hence $F'\left(a\right) = x^T \left[h_1, \ldots, h_k\right] x,$

$$\|F'\left(a\right)\|_\infty = \sum_{k=1}^{n} \left|x^T h_k x\right|,$$

$$\|F'(a) D_a\|_\infty = \sum_{k=1}^n \left|a_k x^T h_k x\right|.$$

It is clear that $\|F'(a) D_a\|_\infty \leq |x^T\| A \| x| \leq \|A\|_\infty$. An open question is therefore to see if for small λ, the structured condition number

$$\mu(G, A) = \frac{1}{\lambda} \sum_{k=1}^n \left|a_k x^T h_k x\right|$$

could be much smaller than $k(G, A) = \frac{1}{\lambda} \|A\|_\infty$.

4. More examples. In this section we list some other matrix maps for which directional derivatives are available and some, other than Toeplitz, linear matrix structures. Structured condition numbers for these maps and matrices can be readily obtained using techniques described above.

4.1 Maps.

i. *Exponential*, $G : A \to e^A$:

$$\frac{\partial G}{\partial h} = \int_0^1 e^{(1-s)A} h e^{sA} ds,$$

see [8], Section 11.3. For example, let A be a symmetric Toeplitz matrix and let $\sigma_1(t), \ldots, \sigma_n(t)$ denote the columns of e^{tA}. Then

$$\|F'(a)\|_\infty = \max_{i,j=1,\ldots,n} \sum_{k=1}^n \left|\int_0^1 \sigma_i^T(1-s) h_k \sigma_j(s) ds\right|,$$

and

$$\|F'(a) D_a\|_\infty = \max_{i,j=1,\ldots,n} \sum_{k=1}^n \left|\int_0^1 \sigma_i^T(1-s) a_k h_k \sigma_j(s) ds\right|,$$

where the matrices $h_1, \ldots, h_k$ are defined as in Section 3.

ii. *Logarithm*, $G : A \to \log(I + A)$:

$$\frac{\partial G}{\partial h} = \int_0^1 (I + sA)^{-1} h (I + sA)^{-1} ds,$$

see Gohberg and Koltracht [5]. Here let $\sigma_i(s), i = 1, \ldots, n$ denote the columns of $(I + sA)^{-1}$. Then

$$\|F'(a)\|_\infty = \max_{i,j=1,\ldots,n} \sum_{k=1}^n \left|\int_0^1 \sigma_i^T(s) h_k \sigma_j(s) ds\right|,$$

and similarly for $\|F'(a) D_a\|_\infty$.

iii. *Full rank least squares,* $Ax_{ls} = b, A$ is $m \times n, m > n,$ rank $A = n, G = [G_1, G_2], h = [E, f]$ as in Section 3.2. Then, see [8] Section 6.1,

$$\frac{\partial G_1}{\partial E} = \left(A^T A\right)^{-1} \left[E^T \left(Ax_{ls} - b\right) + A^T E x_{ls}\right],$$

$$\frac{\partial G_2}{\partial f} = \left(A^T A\right)^{-1} A^T f.$$

iv. *Full rank underdetermined system,* $Ax_{mn} = b, m \leq n,$ rank $A = m, x_{mn}$ is the minimal norm solution. Again, $G = [G_1, G_2], G : [A, b] \to x_{mn}$. Then, see [8] Section 6.7,

$$\frac{\partial G_1}{\partial E} = \left[E^T - A^T \left(AA^T\right)^{-1} \left[AE^T + EA^T\right]\right] \left(AA^T\right)^{-1} b,$$

$$\frac{\partial G_2}{\partial f} = A^T \left(AA^T\right)^{-1} f.$$

v. *Eigenvector of an $n \times n$ matrix with n different eigenvalues $\lambda_1, \ldots, \lambda_n$ and corresponding right $x_1, \ldots, x_n$ and left $y_1, \ldots, y_n$ eigenvectors, $G : A \to x_k$,* eigenvector number k. Then, see [8] Section 7.2,

$$\frac{\partial G}{\partial h} = \sum_{\substack{i=1 \\ i \neq k}}^{n} \frac{y_i^H h x_k}{\left(\lambda_k - \lambda_i\right) y_i^H x_i} x_i ,$$

where H denotes hermitian transposed.

Norms of the derivatives in iii) - v) can be expressed in the same way as in i), ii).

4.2. Linear structures.

i. *Hankel matrices*:

$$A = \begin{bmatrix} a_1 & a_2 & a_3 & \cdot & a_n \\ a_2 & a_3 & \cdot & \cdot & a_{n+1} \\ a_3 & \cdot & \cdot & \cdot & \cdot \\ \cdot & \cdot & \cdot & \cdot & \cdot \\ a_n & a_{n+1} & \cdot & \cdot & a_{2n-1} \end{bmatrix}.$$

As in the Toeplitz case the formulas (3.1) and (3.2) etc. apply with the only difference that the summation is from 1 to $2n - 1$. Here, (apart from the example of the Hilbert matrix reported in the introduction) we do not have much evidence about the relationship between usual and structured condition numbers for the inversion of Hankel matrices.

ii. *Circulant matrices*:

$$A = \begin{bmatrix} a_1 & a_2 & \cdot & \cdot & a_n \\ a_n & a_1 & a_2 & \cdot & a_{n-1} \\ a_{n-1} & a_n & a_1 & a_2 & \cdot \\ \cdot & \cdot & \cdot & \cdot & \cdot \\ a_2 & a_3 & \cdot & a_n & a_1 \end{bmatrix}$$

iii. Brownian matrices

$$
A = \begin{bmatrix}
a_1 & a_2 & a_3 & \cdot & a_n \\
a_2 & a_2 & a_3 & \cdot & a_n \\
a_3 & a_3 & a_3 & \cdot & \cdot \\
\cdot & \cdot & \cdot & \cdot & a_n \\
a_n & a_n & \cdot & a_n & a_n
\end{bmatrix}
$$

iv. *Matrices with a fixed sparseness pattern.* Their structured condition number is given, however, by (2.2).

v. *Block matrices.* All of the above with entries a_k replaced by matrices. These matrices can be structured themselves, e.g. Toeplitz block-Toeplitz matrices.

vi. *Linear combinations of the above,* e.g. Toeplitz plus Hankel, Toeplitz plus diagonal, etc.

vii. Additional examples can be found in Van Loan [14].

REFERENCES

[1] J.J. Dongarra, J.R. Bunch, C.B. Moller and G.W. Stewart, *LINPACK Users Guide*, SIAM Publications, Philadelphia, 1979.

[2] I.Gohberg, T. Kailath, I. Koltracht and P. Lancaster, *Linear complexity parallel algorithms for linear systems of equations with recursive structure*, Linear Alg. and It's Appl., 88/89 (1987) pp. 271-315.

[3] I. Gohberg and I. Koltracht, *On the Inversion of Cauchy Matrices*, Proceedings of the International Symposium MTNS-89, Vol. III, Birkhäuser (1990), pp. 381-392.

[4] I. Gohberg and I. Koltracht, *Componentwise, Mixed and Structured Condition Numbers*, SIAM J. Matrix Anal. Appl. V. 14, No. 3 (July 1993), pp. 688–704.

[5] I. Gohberg and I. Koltracht, *Condition Numbers of Matrix Functions*, Applied Numerical Mathematics 12 (1993), 107–117.

[6] I. Gohberg, I. Koltracht and D. Xiao, *On the Solution of Yule-Walker Equations*, SPIE Proceeding on Advanced Signal Processing Algorithms, Architectures and Implementations II, Vol. 1566 (1991), pp. 14-22.

[7] I. Gohberg, I. Koltracht and D. Xiao, *On Computation of Schur Coefficients of Toeplitz Matrices* (to appear in SIAM J. Matrix Anal. Appl.).

[8] G.H. Golub and C.F. Van Loan, *Matrix Computations*, Second Edition, The John Hopkins University Press, 1989.

[9] D.J. Higham and N.J. Higham, *Backward Error and Condition of Structured Linear Systems*, University of Manchester/SIMAX, Vol. 13, No. 1 (1992), pp. 162-175.

[10] T. Kailath *A View of Three Decades of Linear Filtering Theory*, IEEE Trans. on Information Theory, Vol. IT-20, No. 2 (1974), pp. 145-181.

[11] I. Koltracht and P. Lancaster, *Condition Numbers of Toeplitz and Block-Toeplitz Matrices*, In I. Schur Methods in Operator Theory and Signal Processing, OT-18, 271-300, Birkhäuser Verlag, 1986.

[12] R.D. Skeel, *Scaling for Numerical Stability in Gaussian Elimination*, J. Assoc. Comput. Mach., 26 (1979), pp. 494-526.

[13] J. Stoer and R. Bulirsch, *Introduction to Numerical Analysis*, Springer Verlag, 1980.

[14] P.M. Van Dooren, *Structured Linear Algebra Problems in Digital Signal Processing*, Proceedings of NATO ASI, Leuven 1988. Springer Verlag, Series F, 1990.

THE CANONICAL CORRELATIONS OF MATRIX PAIRS AND THEIR NUMERICAL COMPUTATION

GENE H. GOLUB* AND HONGYUAN ZHA[†]

Abstract. This paper is concerned with the analysis of canonical correlations of matrix pairs and their numerical computation. We first develop a decomposition theorem for matrix pairs having the same number of rows which explicitly exhibits the canonical correlations. We then present a perturbation analysis of the canonical correlations, which compares favorably with the classical first order perturbation analysis. Then we propose several numerical algorithms for computing the canonical correlations of general matrix pairs; emphasis is placed on the case of large sparse or structured matrices.

Key words. canonical correlation, singular value decomposition, perturbation analysis, large sparse matrix, structured matrix

AMS(MOS) subject classifications. primary 15A18, 15A21, 65F15; secondary 62H20

1. Introduction. Given two vectors $u \in \mathcal{R}^n$ and $v \in \mathcal{R}^n$, a natural way to measure the closeness of two *one dimensional* linear subspaces spanned by u and v respectively, is to consider the acute angle formed by the two vectors, the cosine of which is given by

$$\sigma(u, v) := \frac{|u^T v|}{\|u\|_2 \|v\|_2}.$$

We observe that $\sigma(u, v) = 0$, when u and v are orthogonal to each other; and $\sigma(u, v) = 1$, when the two linear subspaces are identical. Given two linear subpaces that are spanned by the columns of matrices $A \in \mathcal{R}^{m \times n}$ and $B \in \mathcal{R}^{m \times l}$, we are concerned with the problem of how to measure the closeness of span$\{A\}$ and span$\{B\}$, the range spaces of A and B. One natural extension of the one dimensional case is to choose a vector from span$\{A\}$, i.e., a linear combination of the columns of A, say Ax, and similarly By from span$\{B\}$, and form $\sigma(By, Ax)$. The closeness of span$\{A\}$ and span$\{B\}$ can be measured by the following

$$d(A, B) = \min_{x \in \mathcal{R}^n,\; y \in \mathcal{R}^l} \sigma(By, Ax).$$

However, the two linear subspaces or rather the matrix pair (A, B) have more structure to reveal than that defined by the minimum. In 1936,

* Computer Science Department, Stanford University, Stanford, CA 94305-2140. email: golub@cholesky.stanford.edu. This work was supported in part by NSF grant DRC-8412314 and Army contract DAAL-03-90-G-0105.

† Scientific Computing & Computational Mathematics, Stanford University, Stanford, CA 94305-2140. email: zha@cholesky.stanford.edu. This work was supported in part by Army contract DAAL-03-90-G-0105. (Current address is: Dept. of Computer Science and Engineering, 307 Pond Laboratory, The Pennsylvania State University, University Park, PA 16802-6103.)

Hotelling proposed to recursively define a sequence of quantities which is now called *canonical correlations* of a matrix pair (A, B) [8].

DEFINITION 1.1. Let $A \in \mathcal{R}^{m \times n}$ and $B \in \mathcal{R}^{m \times l}$, and assume that

$$p = \text{rank}(A) \geq \text{rank}(B) = q.$$

The canonical correlations $\sigma_1(A, B), \cdots, \sigma_q(A, B)$ of the matrix pair (A, B) are defined recursively by the formulae

$$\sigma_k(A, B) = \max_{\substack{Ax \neq 0, \quad By \neq 0, \\ Ax \perp \{Ax_1, \cdots, Ax_{k-1}\}, \\ By \perp \{By_1, \cdots, By_{k-1}\}.}} \sigma(By, Ax) =: \sigma(By_k, Ax_k),$$
$$(1.1)$$
$$k = 1, \cdots, q.$$

It is readily seen that

$$\sigma_1(A, B) \geq \cdots \geq \sigma_q(A, B),$$

and

$$d(A, B) = \sigma_q(A, B).$$

The unit vectors

$$Ax_i/\|Ax_i\|_2, \quad By_i/\|By_i\|_2, \quad (i = 1, \cdots, q)$$

in (1.1) are called the canonical vectors of (A, B); and

$$x_i/\|Ax_i, \quad y_i/\|By_i\|_2, \quad i = 1, \cdots, q$$

are called the canonical weights. Sometimes the angles $\theta_k \in [0, \pi/2]$ satisfying $\cos \theta_k = \sigma_k(A, B)$ are called the principal angles between span$\{A\}$ and span$\{B\}$ [7].[1] The basis of span$\{A\}$ or span$\{B\}$ that consists of the canonical vectors are called the canonical basis.

There are various ways of formulating the canonical correlations, which are all equivalent. They shed insights on the problem from different perspectives, and as we will see later, some of the formulations are more suitable for numerical computation than others. The applications of the canonical correlations are enormous such as system identification, information retrieval, statistics, econometrics, psychology, educational research, anthropology and botany [1] [17] [9]. There are also many variants and generalizations of the canonical correlations: to the case of more than two matrices (surveyed by Kettenring [11], see also [17]); to sets of random functions [2]; to nonlinear transformations [17]; and to problems with

---

[1] As is pointed by G.W. Stewart [15], the concept of canonical angles between two linear subspaces is much older than canonical correlations, and can be traced back to C. Jordan [10, p.129 Equation(60)].

(in)equality constraints. Several numerical algorithms have been proposed for the computation of the canonical correlations and the corresponding canonical vectors (see Björck and Golub's paper [4] and references therein); however, in the literature there is very little discussion of the case of large sparse and structured matrix pairs, which will receive a fairly detailed treatment in Section 4.

The organization of the paper is as follows: in Section 2, we present several different formulations of the canonical correlations; in Section 3, we develop a decomposition theorem for general matrix pairs having the same number of rows: this decomposition not only explicitly exhibits the canonical correlations of the matrix pair, it also reveals some of its other intrinsic structures. We also discuss the relation between the canonical correlations and the corresponding eigenvalue problem and the RSVD [20]. In Section 4, we present perturbation analyses of the canonical correlations; the results compare favorably with the classical first order counterpart developed in [4]. We derive perturbation bounds for the normwise as well as componentwise perturbations. In Section 5, we propose several numerical algorithms for computing the canonical correlations. For the case of dense matrices, we also discuss the updating problem. The emphasis of the section is placed on the case of large sparse or structured matrix pairs. We will first present an algorithm using alternating linear least squares approach which has a nice geometric interpretation. We also relate this algorithm to a modified power method and derive its convergence rate. Then we adapt the Lanczos bidiagonalization process to compute a few of the largest canonical correlations. Our algorithms have the attractive feature that it is not necessary to compute the orthonormal basis of the column space of A and B as is used in Björck-Golub's algorithm, and thus one can fully take advantage of the sparsity or special structures (e.g, Hankel or Toeplitz structures) of the underlying matrices. Numerical examples will also be given to illustrate the algorithms.

2. Several different formulations. There are quite a few different ways of defining and formulating canonical correlations: Hotelling's original derivation is based on matrix algebra and analysis [8]; Rao and Yanai used the theory of orthogonal projectors [14]; Escoufier proposed a general frame work for handling data matrix by matrix operators, which also includes the canonical correlations as a special case [6]; Björck and Golub used matrix decomposition of the given data matrices [4]. In this section, we give some of the formulations and indicate their equivalence.

The Singular Value Decomposition (SVD) Formulation. Let the QR decomposition of A and B be

$$A = Q_A R_A, \quad B = Q_B R_B,$$

where Q_A and Q_B are orthonormal matrices, and R_A and R_B are nonsingular upper triangular matrices, then

$$\sigma(By, Ax) = \frac{y^T B^T Ax}{\|By\|_2 \|Ax\|_2} = \frac{y^T R_B^T Q_B^T Q_A R_A x}{\|R_B y\|_2 \|R_A x\|_2} =: v^T Q_B^T Q_A u,$$

where we have designated $u = R_A x / \|R_A x\|_2$ and $v = R_B y / \|R_B y\|_2$. Using a characterization of the SVD [7, p.428], we see that the canonical correlations are the singular values of $Q_B^T Q_A$, and if

$$Q_B^T Q_A = P^T \operatorname{diag}(\sigma_1(A, B), \cdots, \sigma_q(A, B)) Q$$

represents the SVD of $Q_B^T Q_A$, then

$$Q_A P(:, 1:q) = [u_1, \cdots, u_q], \quad \text{and} \quad Q_B Q = [v_1, \cdots, v_q]$$

give the canonical vectors of (A, B). Note that since $Q_B^T Q_A$ is a section of an orthogonal matrix, $\sigma_k(A, B) \leq 1$, $k = 1, \cdots, q$. We also note that the canonical vectors are not unique if, say $\sigma_k(A, B) = \sigma_{k+1}(A, B)$. However, the above formulation is rather general in the sense that it can also handle the case when A and/or B are rank deficient.

A Trace Maximization Formulation. Let us consider the following maximization problem:

$$(2.1) \qquad \max_{\substack{L^T B^T BL = I_p \\ M^T A^T AM = I_p}} \operatorname{trace}(L^T B^T AM),$$

where for simplicity we have further assumed that $p = q$; otherwise we can append zero columns to B to make the pair (A, B) satisfy this assumption. Again using the QR decomposition of A and B, we see that the two equality constraints in (2.1) imply that $R_A M$ and $R_B L$ are orthogonal matrices, and we arrive at the following equivalent maximization problem

$$(2.2) \qquad \max_{U \text{ and } V \text{ are orthogonal}} \operatorname{trace}(U^T (Q_B^T Q_A) V).$$

To this end, we cite a well-known result of Von Neumann [16].

LEMMA 2.1. *Let the singular values of A and B be*

$$\sigma_1 \geq \sigma_2 \geq \cdots \geq \sigma_n \text{ and } \tau_1 \geq \tau_2 \geq \cdots \geq \tau_n.$$

Then

$$\max_{U \text{ and } V \text{ are orthogonal}} \operatorname{trace}(BU^T AV) = \Sigma_i \sigma_i \tau_i.$$

The above problem (2.2) is a special case of the lemma by choosing $B = I$ and $A = Q_B^T Q_A$.

Remark 2.1. Since $L^T B^T BL = I_p$, $M^T A^T AM = I_p$, the maximization problem (2.1) is equivalent to the following minimization problem:

$$(2.3) \qquad \min_{\substack{L^T B^T BL = I_p \\ M^T A^T AM = I_p}} \|AM - BL\|_F,$$

which can be interpreted as finding an orthonormal basis of span$\{A\}$ and span$\{B\}$ respectively, such that their difference measured in the Frobenius norm are minimized. It is equivalent to the following orthogonal *Procrustes problem.* Let Q_A and Q_B be any orthonormal basis of span$\{A\}$ and span$\{B\}$, respectively. Then (2.1) is equivalent to

$$\min_{U \text{ is orthogonal}} \|Q_A - Q_B U\|_F.$$

We note that the above is a special Procrustes problem where Q_A and Q_B are orthonormal, while in the general case, Q_A and Q_B can be replaced by two general matrices [7, Section 12.4.1].

A Lagrange Multiplier Formulation [8]. For the constrained minimization problem (1.1), write the Lagrange multiplier function

$$f(x, y, \lambda, \mu) = y^T B^T Ax - \lambda(\|Ax\|_2^2 - 1) - \mu(\|By\|_2^2 - 1).$$

Differentiating with respect to x, y, λ, and μ leads to:

$$(2.4) \qquad \begin{aligned} B^T Ax - \mu B^T By &= 0, \\ A^T By - \lambda A^T Ax &= 0, \\ y^T B^T By &= 1, \\ x^T A^T Ax &= 1. \end{aligned}$$

It follows that $\lambda = \mu$ and

$$\begin{pmatrix} O & B^T A \\ A^T B & O \end{pmatrix} \begin{pmatrix} y \\ x \end{pmatrix} = \lambda \begin{pmatrix} B^T B & O \\ O & A^T A \end{pmatrix} \begin{pmatrix} y \\ x \end{pmatrix}.$$

Therefore finding the canonical correlations, which are the stationary values, corresponds to solving for the eigenvalues of the above generalized eigenvalue problem. On the other hand, since

$$\left\| \frac{Ax}{\|Ax\|_2} - \frac{By}{\|By\|_2} \right\|_2^2 = 2\left(1 - \frac{y^T B^T Ax}{\|By\|_2 \|Ax\|_2}\right),$$

the first canonical correlation can also be computed by solving the minimization problem

$$\min_{Ax,\, By \neq 0} \left\| \frac{Ax}{\|Ax\|_2} - \frac{By}{\|By\|_2} \right\|_2.$$

One way of solving the minimization problem is to first fix y, and find the optimal x; then fix x at this optimal value, and then solve for y and so on. At each iteration step, we can reformulate the problem as

$$\min_{\text{subject to } \|Az\|_2=1} \|w - Az\|_2 \,,$$

where w is of unit length. Using the Lagrange multiplier method, we seek to minimize

$$f(z, \lambda) = \|w - Az\|_2^2 + \lambda(\|Az\|_2^2 - 1) \,.$$

Writing down the first order condition for the stationary values, we obtain

$$A^T Az = A^T w/(1 + \lambda), \quad z^T A^T Az = 1 \,;$$

and the solution is given by

$$\lambda = (A^T w)^T (A^T A)^{-1} (A^T w) - 1 \quad z = (A^T A)^{-1} A^T w/(1 + \lambda)$$
$$= w^T P_A w - 1, \qquad\qquad\qquad = A^\dagger w/(1 + \lambda) \,,$$

where $P_A^2 = P_A$ is the orthogonal projection onto $\text{span}\{A\}$. We note that z is in the direction of $A^\dagger w$, and is the least squares solution of

$$\min_x \|w - Az\|_2.$$

Actually, this approach will lead to the alternating least squares (ALS) method that we will discuss in Section 5.

3. A decomposition theorem. It is readily checked from Definition 1 that the canonical correlations are invariant under the following group transformation

$$A \longrightarrow QAX_A, \quad B \longrightarrow QBX_B \,,$$

where Q is orthogonal and X_A and X_B are nonsingular. The following theorem gives the maximum invariants of a matrix pair (A, B) under the above group transformation. It also provides information on other structures of the matrix pair as well. It can be considered as a recast of Theorem 5.2 in [16, pp. 40-42] (cf. [4, Equation (15)] [18, Equation (2.2)]).

THEOREM 3.1. *Let $A \in R^{m \times n}$ and $B \in R^{m \times l}$, and assume that*

$$p = \text{rank}(A) \geq \text{rank}(B) = q \,.$$

Then there exists orthogonal matrix Q and nonsingular matrices X_A and X_B such that

$$A = Q[\Sigma_A, O]X_A, \quad B = Q[\Sigma_B, O]X_B \,,$$

where $\Sigma_A \in R^{m \times p}$ and $\Sigma_B \in R^{m \times q}$ are of the following form

$$(3.1) \qquad \Sigma_A = \begin{pmatrix} \begin{array}{c|c} \begin{matrix} I_i & & \\ & C & \\ & & O \end{matrix} & \\ \hline O & \begin{matrix} S & \\ & I_k \end{matrix} \end{array} \end{pmatrix}, \quad \Sigma_B = \begin{pmatrix} I_q \\ \hline O \end{pmatrix},$$

with

$$(3.2) \qquad \begin{aligned} C &= \mathrm{diag}(\alpha_{i+1} \cdots \alpha_{i+j}), \ 1 > \alpha_{i+1} \geq \cdots \geq \alpha_{i+j} > 0, \\ S &= \mathrm{diag}(\beta_{i+1}, \cdots, \beta_{i+j}), \ 0 < \beta_{i+1} \leq \cdots \leq \beta_{i+j} < 1, \\ & \alpha_{i+1}^2 + \beta_{i+1}^2 = 1, \cdots, \alpha_{i+j}^2 + \beta_{i+j}^2 = 1, \end{aligned}$$

and $p = i + j + k$. The canonical correlations of (A, B) are the diagonal elements of $\Sigma := \mathrm{diag}(I_i, C, O)$. Moreover, we have

$$(3.3) \qquad \begin{aligned} i &= \mathrm{rank}(A) + \mathrm{rank}(B) - \mathrm{rank}([A, B]), \\ j &= \mathrm{rank}([A, B]) + \mathrm{rank}(B^T A) - \mathrm{rank}(A) - \mathrm{rank}(B), \\ k &= \mathrm{rank}(A) - \mathrm{rank}(B^T A). \end{aligned}$$

Proof. Using the QR decomposition, we can transform A and B to

$$A = [Q_A, O]R_A, \quad B = [Q_B, O]R_B,$$

where $Q_A \in R^{m \times p}$ and $Q_B \in R^{m \times q}$ are orthonormal, and R_A and R_B are nonsingular. We then find U orthogonal such that

$$Q_B = U \begin{pmatrix} I_q \\ \hline O \end{pmatrix}.$$

We partition $U^T Q_A$ as $U^T Q_A = [A_1^T, A_2^T]^T$ with $A_1 \in R^{q \times q}$. Let the SVD of A_1 be

$$A_1 = U_1 \begin{pmatrix} I_i & & \\ & C & \\ & & O \end{pmatrix} V_2^T \, ;$$

C defined as in 3.2, and $A_2 V_2 = [A_{11}, A_{12}, A_{13}]$ be partitioned compatibly.[2] Then we have $A_{11} = O$ and A_{13} is orthonormal and can be written as $A_{13} = U_2[O, I_k]^T$ with U_2 an orthogonal matrix. Hence $U_2^T A_{12} = [\tilde{A}_{12}^T, O]^T$, with its last k rows equal zero. Since the columns of $\tilde{A}_{12}$ are orthogonal to each

---

[2] Since A_1 is a section of an orthogonal matrix, all its singular values are less than or equal to one.

other, we can find an orthogonal matrix U_3 such that $\tilde{A}_{12} = U_3[O, S]^T$. The relations in 3.2 follow from the fact that $U^T Q_A$ is orthonormal. Accumulating all the transformations establishes the decomposition (3.1). For the rank expressions of the integer indices we observe that

$$\text{rank}(A) = i+j+k, \text{rank}(B) = q, \text{rank}(B^T A) = i+j, \text{rank}([A, B]) = j+k+q.$$

Some elementary calculation leads to the result (3.3). $\qquad\square$

Remark 3.1. The dimension of $\mathcal{R}(A) \cap \mathcal{R}(B)$ is exactly the number of those canonical correlations of (A, B) which are equal to one.

COROLLARY 3.2. *Let* $Q = (Q_1, Q_2, Q_3, Q_4, Q_5, Q_6)$ *be compatibly partitioned with the block row partitioning of* Σ_A, *i.e.,* $Q_1 \in R^{m \times i}$, $Q_2 \in R^{m \times j}$ *and so on. Then*

$$\text{span}\{Q_1, Q_2 C + Q_5 S, Q_6\} = \mathcal{R}(\mathcal{A}) \,;$$
$$\text{span}\{Q_1, Q_2, Q_3\} = \mathcal{R}(\mathcal{B}) \,;$$
$$\text{span}\{Q_3, -Q_2 S + Q_5 C, Q_4\} = \mathcal{R}(\mathcal{A})^{\perp} \,;$$
$$\text{span}\{Q_4, Q_5, Q_6\} = \mathcal{R}(\mathcal{B})^{\perp} \,;$$
$$\text{span}\{Q_1\} = \mathcal{R}(\mathcal{A}) \cap \mathcal{R}(\mathcal{B}) \,;$$
$$\text{span}\{Q_3\} = \mathcal{R}(\mathcal{A})^{\perp} \cap \mathcal{R}(\mathcal{B}) \,;$$
$$\text{span}\{Q_4\} = \mathcal{R}(\mathcal{A})^{\perp} \cap \mathcal{R}(\mathcal{B})^{\perp} \,;$$
$$\text{span}\{Q_6\} = \mathcal{R}(\mathcal{A}) \cap \mathcal{R}(\mathcal{B})^{\perp} \,.$$

Proof. We prove $\text{span}\{Q_3\} = \mathcal{R}(A)^{\perp} \cap \mathcal{R}(B)$; the other formulae can be similarly established. It is easy to see that $\text{span}\{Q_3\} \subseteq \mathcal{R}(A)^{\perp} \cap \mathcal{R}(B)$; However

$$G := (Q_1, Q_2, Q_3)^T (Q_3, -Q_2 S + Q_5 C, Q_4) = \begin{pmatrix} O & O & O \\ O & -S & O \\ I_k & O & O \end{pmatrix}.$$

It follows that the number of singular values of G that are equal to one is exactly the column dimension of Q_3; the result follows from the comment in Remark 3.1. $\qquad\square$

COROLLARY 3.3. *We also have the following expressions for the dimensions of some of the linear subspaces in Corollary 3.2:*

$$\dim(\mathcal{R}(\mathcal{A}) \cap \mathcal{R}(\mathcal{B})) = \text{rank}(A) + \text{rank}(B) - \text{rank}([A, B]) \,;$$
$$\dim(\mathcal{R}(\mathcal{A})^{\perp} \cap \mathcal{R}(\mathcal{B})) = \text{rank}(B) - \text{rank}(B^T A) \,;$$
$$\dim(\mathcal{R}(\mathcal{A})^{\perp} \cap \mathcal{R}(\mathcal{B})^{\perp}) = m - \text{rank}([A, B]) \,;$$
$$\dim(\mathcal{R}(\mathcal{A}) \cap \mathcal{R}(\mathcal{B})^{\perp}) = \text{rank}(A) - \text{rank}([A, B]) \,.$$

Proof. All the expressions can be proved by using the above corollary and the rank formulae in Theorem 3.1. $\qquad\square$

COROLLARY 3.4. *The finite and zero eigenvalues of*

$$\begin{pmatrix} O & B^T A \\ A^T B & O \end{pmatrix} x = \lambda \begin{pmatrix} B^T B & O \\ O & A^T A \end{pmatrix} x$$

are $\pm\sigma_1(A, B), \cdots, \pm\sigma_q(A, B)$. *And if* $B^T A = I$, *then the nonzero singular values of the matrix product* AB^T *are* $1/\sigma_1(A, B), \cdots, 1/\sigma_q(A, B)$.[3]

Proof. The result can be proved by using the decomposition in Theorem 3.1 and direct computation. It follows that the canonical correlations can also be found by the RSVD of the matrix triplet $(B^T A, B^T, A)$ [20, p.193]; if $B^T A = I$, the RSVD of $(B^T A, B^T, A)$ reduces to the PSVD of (B^T, A) [20, Corollary 4.2]. $\quad\square$

4. Perturbation analyses. In this section we establish some perturbation bounds for the canonical correlations; some of the techniques used here are first devised by Paige in his analysis of the generalized singular value decomposition [12]. We also mention that Björck and Golub developed a first order perturbation analysis in their paper [4]. Before we discuss the general case, let us first consider a simple example:

Example 4.1. We consider the matrix pair:

$$A = \begin{pmatrix} 0 \\ 1 \\ 0 \end{pmatrix}, \quad B = \begin{pmatrix} 1 & 1 \\ 0 & \epsilon \\ 1 & 1 \end{pmatrix},$$

where ϵ is a small quantity. Since $A = (B(:, 2) - B(:, 1))/\epsilon$, hence $\sigma(A, B) = 1$. But if we perturb B to

$$\tilde{B} = \begin{pmatrix} 1 & 1 - \epsilon \\ 0 & 0 \\ 1 & 1 \end{pmatrix} = B - \begin{pmatrix} 0 & \epsilon \\ 0 & \epsilon \\ 0 & 0 \end{pmatrix},$$

since A is orthogonal to the columns of $\tilde{B}$, we have $\sigma(A, \tilde{B}) = 0$. Therefore a small change in the matrix pair (A, B) causes a large change of its canonical correlations. We note that B and $\tilde{B}$ are of the same rank.

We observe that $\text{cond}(B) \approx 1/\epsilon$. This example suggests that the canonical correlations are sensitive to perturbations if the condition number of A or B is large. Now we turn to the discussion of the general case. Using the QR decomposition with column pivoting, A and B can be factorized as

$$A = Q_A R_A, \quad B = Q_B R_B,$$

where Q_A and Q_B are orthonormal, and R_A and R_B are of full row rank. The canonical correlations are simply the singular values of $Q_A^T Q_B$ (cf. [4]).

---

[3] The first result is proved in [8], and the second is also implicit in [20].

Let the SVD of $Q_A^T Q_B$ be

$$Q_A^T Q_B = U\Sigma V^T.$$

We denote the perturbed quantities by adding "$\sim$" to the corresponding unperturbed ones. We assume that A and $\tilde{A}$, B and $\tilde{B}$ are of the same rank.

Let the orthogonal complement of Q_A and Q_B be denoted by $\hat{Q}_A$ and $\hat{Q}_B$ respectively. We define

$$\delta_2(A) = \min_{U \text{ is orthogonal}} \|Q_A - Q_{\tilde{A}} U\|_2 \,,$$

$$\delta_F(A) = \min_{U \text{ is orthogonal}} \|Q_A - Q_{\tilde{A}} U\|_F \,.$$

We note that $\delta_F(A)$ is introduced in [12], and actually it is a special case of the following well known *Procrustes problem* [7]:

$$\min_{Q \text{ is orthogonal}} \|A - BQ\|_F \,,$$

where A and B are arbitrary matrices with same number of columns and rows. The solution of the Procrustes problem can be obtained using the SVD of $B^T A$: let $B^T A = U\Sigma V^T$, then the optimal Q is given by $Q = UV^T$ [7]. For $\delta_F(A)$, some interesting relations can be derived by invoking the CS-decomposition of $(Q_A, \hat{Q}_A)^T \tilde{Q}_A$ [12] [16]:

$$\begin{pmatrix} Q_A^T Q_{\tilde{A}} \\ \hat{Q}_A^T Q_{\tilde{A}} \end{pmatrix} = \begin{pmatrix} U_A C_A W_A^T \\ V_A S_A W_A^T \end{pmatrix},$$

where U_A, V_A and W_A are orthogonal, and C_A and S_A are quasi-diagonal. Then we have [12]:

$$\delta_F(A)^2 = 2\Sigma_i\,(1 - \sigma_i) \leq 2\Sigma_i\,(1 - \sigma_i^2)$$

(4.1)
$$= 2\|S\|_F^2 = 2\|\hat{Q}_A^T Q_{\tilde{A}}\|_F^2$$

$$= 2\Sigma_i\,(1 - \sigma_1)(1 + \sigma_1) \leq 2\delta_F(A)^2 \,,$$

where we used $C_A = \text{diag}(\sigma_1, \cdots, \sigma_q)$ with $\sigma_1 \leq \cdots \leq \sigma_q$.

For the $\delta_2(A)$, we proceed as follows:

$$\|Q_A - Q_{\tilde{A}} U\|_2^2 = \sigma_{\max}(2I + Q_A^T Q_{\tilde{A}}(-U) + (Q_A^T Q_{\tilde{A}}(-U))^T)$$

$$\geq 2 - \sigma_{\max}(Q_A^T Q_{\tilde{A}}(-U) + (Q_A^T Q_{\tilde{A}}(-U))^T)$$

(4.2)
$$\geq 2 - 2\sigma_{\max}(Q_A^T Q_{\tilde{A}}(-U))$$

$$= 2 - 2\sigma_{\max}(Q_A^T Q_{\tilde{A}})$$

$$= 2(1 - \sigma_q) \,.$$

Hence $\delta_2(A) \geq \sqrt{2(1 - \sigma_q)}$. However, by choosing the canonical basis in the CS-decomposition we have

$$
\text{(4.3)} \qquad
\begin{aligned}
\delta_2(A) &\leq \left\| \begin{pmatrix} I \\ O \\ O \end{pmatrix} - \begin{pmatrix} C_A \\ S_A \\ O \end{pmatrix} \right\|_2 \\
&= \sqrt{\lambda_{\max}((I - C_A)^2 + S_A^2)} \\
&= \sqrt{2(1 - \sigma_1)}.
\end{aligned}
$$

Therefore we obtain

$$
\sqrt{2(1 - \sigma_q)} \leq \delta_2(A) \leq \sqrt{2(1 - \sigma_1)}.
$$

It follows that

$$
\text{(4.4)} \qquad
\begin{aligned}
\delta_2(A)^2 &\leq 2(1 - \sigma_1) \leq 2(1 - \sigma_1^2) \\
&= 2\|S\|_2^2 = 2\|\hat{Q}_A^T Q_{\tilde{A}}\|_2^2.
\end{aligned}
$$

We should remark here that, there is generally no closed form solution to the following *Procrustes* problem:

$$
\min_{Q \text{ is orthogonal}} \|A - BQ\|_2,
$$

when A and B are orthonormal. With the above preparation, we are now ready to prove the following perturbation theorems.

THEOREM 4.1. *Let A and $\tilde{A}$, and B and $\tilde{B}$ have the same rank, and let the condition numbers of A and B be*

$$
\kappa(A) = \|A\|_2\|A^{\dagger}\|_2, \quad \kappa(B) = \|B\|_2\|B^{\dagger}\|_2.
$$

Then

$$
\|\Sigma - \tilde{\Sigma}\|_2 \leq \sqrt{2}\left\{ \kappa(A)\frac{\|A - \tilde{A}\|_2}{\|A\|_2} + \kappa(B)\frac{\|B - \tilde{B}\|_2}{\|B\|_2} \right\}.
$$

Proof. For any orthogonal matrices U and V, we have

$$
\text{(4.5)} \qquad
\begin{aligned}
\|\Sigma - \tilde{\Sigma}\|_2 &\leq \|Q_A^T Q_B - U^T Q_{\tilde{A}}^T Q_{\tilde{B}} V\|_2 \\
&= \|(Q_A - Q_{\tilde{A}}U)^T Q_B + U^T Q_{\tilde{A}}^T(Q_B - Q_{\tilde{B}}V)\|_2 \\
&\leq \|Q_A - Q_{\tilde{A}}U\|_2 + \|Q_B - Q_{\tilde{B}}V\|_2.
\end{aligned}
$$

Since U and V are arbitrary, we obtain

$$
\begin{aligned}
\|\Sigma - \tilde{\Sigma}\|_2 &\leq \delta_2(A) + \delta_2(B) \\
&\leq \sqrt{2}(\|\hat{Q}_A^T Q_{\tilde{A}}\|_2 + \|\hat{Q}_B^T Q_{\tilde{B}}\|_2) \,.
\end{aligned}
\tag{4.6}
$$

Let $\tilde{A} = A + \Delta A$, then

$$
\hat{Q}_{\tilde{A}}^T \Delta A = -\hat{Q}_{\tilde{A}}^T Q_A R_A, \quad \hat{Q}_A^T \Delta A = \hat{Q}_A^T Q_{\tilde{A}} R_{\tilde{A}} \,.
\tag{4.7}
$$

Since R_A and $R_{\tilde{A}}$ are of full row rank:

$$
\hat{Q}_{\tilde{A}}^T Q_A = -\hat{Q}_{\tilde{A}}^T \Delta A R_A^\dagger, \quad \hat{Q}_A^T \tilde{Q}_A = \hat{Q}_A^T \Delta A R_{\tilde{A}}^\dagger.
$$

Therefore

$$
\|\hat{Q}_{\tilde{A}}^T Q_A\|_2 = \|\hat{Q}_A^T \tilde{Q}_A\|_2 \leq \|\Delta A\|_2 \min\{\|A^\dagger\|_2, \|\tilde{A}^\dagger\|_2\} \,.
$$

We can also establish similar results for B; therefore

$$
\begin{aligned}
\|\Sigma - \tilde{\Sigma}\|_2 &\leq \sqrt{2}(\|A - \tilde{A}\|_2 \min\{\|A^\dagger\|_2, \|\tilde{A}^\dagger\|_2\} \\
&\qquad + \|B - \tilde{B}\|_2 \min\{\|B^\dagger\|_2, \|\tilde{B}^\dagger\|_2\}) \\
&\leq \sqrt{2}\{\kappa(A)\|A - \tilde{A}\|_2/\|A\|_2 + \kappa(B)\|B - \tilde{B}\|_2/\|B\|_2\} \,,
\end{aligned}
\tag{4.8}
$$

which establishes the result. $\quad\square$

Using the same technique, we can also prove the following result for the case of Frobenius norm:

THEOREM 4.2. *Let A and $\tilde{A}$, and B and $\tilde{B}$ have the same rank, then*

$$
\begin{aligned}
\|\Sigma - \tilde{\Sigma}\|_F &\leq \sqrt{2}(\|A - \tilde{A}\|_F \min\{\|A^\dagger\|_2, \|\tilde{A}^\dagger\|_2\} \\
&\quad + \|B - \tilde{B}\|_F \min\{\|B^\dagger\|_2, \|\tilde{B}^\dagger\|_2\})
\end{aligned}
$$

Remark 4.1. In [4], Björck and Golub derived the following first order perturbation bound:

$$
\|\Sigma - \tilde{\Sigma}\|_2 \leq \epsilon_A \sin \theta_{\max}(A, \tilde{B}) + \epsilon_B \sin \theta_{\max}(A, B) + O(\delta^2) \,,
$$

where

$$
\|A - \tilde{A}\|_2 \leq \epsilon_A \|A\|_2, \quad \|B - \tilde{B}\|_2 \leq \epsilon_B \|B\|_2, \quad \delta = \epsilon_A \kappa(A) + \epsilon_B \kappa(B) \,.
$$

Our result (Theorem 4.1) compares favorably to the above.

The above theorems well the perturbation result in Example 4.1, but they do not tell the whole story as is demonstrated by the following example. First, let us consider the following matrix pair

Example 4.2. Given the matrix pair,

$$A = \begin{pmatrix} 1 \\ 0 \\ -1 \end{pmatrix}, \quad B = \begin{pmatrix} 1 & 1 \\ 0 & 10^{-10} \\ 1 & 1 \end{pmatrix}.$$

The computed Q in the QR decomposition of B is

$$Q =$$

$$\begin{array}{rr} -0.70710678118655 & 0.00000125385069 \\ 0 & -0.99999999999843 \\ -0.70710678118655 & -0.00000125385069 \end{array}$$

and the computed canonical correlation is

$$1.773212653097254e - 06.$$

All the computation in this section was carried out on a Sun 3/50 workstation using MATLAB version 3.5e with machine precision eps$\approx$ 2.22e-16. Since cond$(B) \approx 10^{10}$, this result coincides with the prediction given by the bounds in Theorem 4.1. Now let us consider another matrix pair,

Example Given the matrix pair

$$A_1 = \begin{pmatrix} 1 \\ 0 \\ -1 \end{pmatrix}, \quad B_1 = \begin{pmatrix} 1 & 10^{10} \\ .4 & .9 \\ 1 & 10^{10} \end{pmatrix}.$$

The matrix Q in the QR decomposition of B_1 is

$$Q =$$

$$\begin{array}{rr} -0.68041381743977 & 0.19245008972988 \\ -0.27216552697591 & -0.96225044864938 \\ -0.68041381743977 & 0.19245008972987 \end{array}.$$

We also compute $\sigma(A, B)$ as

$$7.654331812476101e - 16.$$

But since cond$(B_1) \approx 10^{10}$, and the machine precision is approximately 10^{-16}, the bounds in the above theorems will predict a perturbation of size

$$\text{cond}(B) \times \text{eps} \approx 10^{-6},$$

which is much larger than the computed result. However, theoretically if we scale the last column of B, we get a well conditioned matrix, and column scaling does not change the canonical correlations. The perturbation bounds in the above theorems are not invariant under the column scaling of A and B, therefore we need a refined version of the perturbation bounds.

Before we proceed, we introduce some more notation. If $A = (a_{ij})$, then we write $|A| := (|a_{ij}|)$; We denote $|A| \le |B|$, if $|a_{ij}| \le |b_{ij}|$; it is easy to verify that if $A = BC$, then $|A| \le |B||C|$.

We define the column-scaling independent condition number of A as

$$\kappa_S(A) = \left\| |R||R^{-1}| \right\|_2,$$

if the QR decomposition of A is $A = QR$. Obviously, $\kappa_S(A)$ is independent of the column scaling of A; i.e.,

$$\kappa_S(AD) = \kappa_S(A)$$

for all positive definite diagonal matrix D.

THEOREM 4.3. *Let A and B be of full column rank, $\tilde{A} = A + \Delta A$ and $\tilde{B} = A + \Delta B$ with $|\Delta A| \le \epsilon|A|$ and $|\Delta B| \le \epsilon|B|$, and $\tilde{A}$ and $\tilde{B}$ are also of full column rank. Then*

$$\|\Sigma - \tilde{\Sigma}\|_2 \le \sqrt{2}\epsilon(\sqrt{p(m-p)}\kappa_S(A) + \sqrt{q(m-q)}\kappa_S(B)).$$

Proof. From (4.7), we have

$$\hat{Q}_{\tilde{A}}^T Q_A = -\hat{Q}_{\tilde{A}}^T \Delta A R_A^{-1}.$$

It follows that

$$|\hat{Q}_{\tilde{A}}^T Q_A| \le |\hat{Q}_{\tilde{A}}^T||\Delta A||R_A^{-1}| \le \epsilon|\hat{Q}_{\tilde{A}}^T||Q_A||R_A| \cdot |R_A^{-1}|.$$

Hence

$$(4.9) \quad \begin{aligned} \|\Sigma - \tilde{\Sigma}\| &\le \sqrt{2}\left(\|\hat{Q}_A^T Q_{\tilde{A}}\|_F + \|\hat{Q}_B^T Q_{\tilde{B}}\|_F\right) \\ &\le \sqrt{2}\,\epsilon\left(\left\| |\hat{Q}_{\tilde{A}}^T||Q_A| \right\|_F \kappa_S(A) + \left\| |\hat{Q}_{\tilde{B}}^T||Q_B| \right\|_F \kappa_S(B)\right), \end{aligned}$$

and the result is established. $\qquad\square$

5. Numerical algorithms. In this section, we discuss numerical computation of the canonical correlations and the corresponding canonical vectors. For simplicity, throughout the section we assume that both A and B are of full column rank, i.e., $A \in R^{m \times p}$ and $B \in R^{m \times q}$, and

$$p = \text{rank}(A) \ge \text{rank}(B) = q.$$

5.1. Dense matrices. The following algorithm based on SVD was proposed by Björck and Golub [4], [7, Chapter 12].

Algorithm 5.1. Given A and B, the following procedure computes the orthonormal matrices $U = [u_1, \cdots, u_q]$ and $V = [v_1, \cdots, v_q]$ and $\sigma_1(A, B)$, $\cdots, \sigma_q(A, B)$ where $\{\sigma_k(A, B)\}$ are the canonical correlations of (A, B) and the u_k and v_k are the associated canonical vectors.

i) Compute the QR decomposition of A and B:

$$A = Q_A R_A, \quad \text{where} \quad Q_A^T Q_A = I_p \,,$$

$$B = Q_B R_B, \quad \text{where} \quad Q_B^T Q_B = I_q \,.$$

ii) Form $C = Q_B^T Q_A$, and compute the SVD of C: $C = P^T$
   diag$(\sigma_i(A, B))Q$.

$$Q_A P(:, 1:q) = [u_1, \cdots, u_q], \quad Q_B Q = [v_1, \cdots, v_q] \,.$$

As discussed in Section 4, there is no need to scale the columns of A and B before we compute their QR decompositions. Some numerical experiments were reported in [4], where QR decomposition with column pivoting was used to handle the rank deficient case.

5.2. Updating problems. Let B be augmented by a column vector b. We want to investigate the relation between the canonical correlations of (A, B) and those of $(A, [B, b])$. We will develop an algorithm for updating the canonical correlations. We summarize the result in the following theorem.

THEOREM 5.1. *Let g be a unit vector that spans $\mathcal{R}([B, b]) \cap \mathcal{R}(B)^\perp$.*[4] *Let Q_s be the orthonormal basis of the subspace $\mathcal{R}(A)^\perp \cap \mathcal{R}(B)^\perp$. Define*

$$\eta = \|(I - Q_s Q_s^T)g\|_2 \,.$$

Then[5]
1. $\sigma_l(A, [B, b]) = 1, \quad l = 1, \cdots, i;$
2. $\sigma_l(A, B) \leq \sigma_l(A, [B, b]) \leq \sigma_l(A, B)\sqrt{1 + (1 - \eta^2)\tan^2\theta_l}\,,$

   $l = i + 1, \cdots, i + j;$ *where θ_l is the l-th canonical (principal) angle (see Definition 1); and*
3. $0 \leq \sigma_{i+j+1}(A, [B, b]) \leq 1 - \eta^2, \sigma_l(A, [B, b]) = 0\,,$

   $l = i + j + 2, \cdots, i + j + k.$

Proof. We consider the case when $g \neq 0$; the other case when $g = 0$ is trivial. Let the QR decomposition of A and B be

$$A = Q_A R_A, \quad B = Q_B R_B \,.$$

---

[4] If $\mathcal{R}([B,b]) \perp \mathcal{R}(B)^\perp$, i.e., $b \in \mathcal{R}(B)$ then we take $g = 0$.
[5] The integer indices refer to those in Theorem 3.1.

Using Theorem 3.1, we write

$$Q_A = Q\Sigma_A U^T, \quad Q_B = Q\Sigma_B V^T$$

where Q, U and V are orthogonal, and Σ_A and Σ_B are given by (3.1). The QR decomposition of $[B, b]$ can be written as

$$[B, b] = [Q_B, g]R_{[B,b]}$$

with $R_{[B,b]}$ a nonsingular upper triangular matrix. Let $Q = (Q_1, Q_2)$ with $Q_1 \in R^{m \times q}$, then there exists a unit vector $\tilde{g} \in R^{m-q}$ such that $g = Q_2\tilde{g}$. Hence $\tilde{g} = Q_2^T g$. Let $\tilde{g}^T = (g_1^T, g_2^T, g_3^T)^T$ with $g_2 \in R^j$ and $g_3 \in R^k$. Using Corollary 3.2, we have $Q_s^T g = g_1$. On the other hand

$$(5.1) \qquad H := \begin{pmatrix} Q_B^T \\ g^T \end{pmatrix} Q_A = \operatorname{diag}\{\tilde{V}, 1\} \begin{pmatrix} I_i & & & \\ & C & & \\ & & & O \\ 0 & g_2^T S & g_3 \end{pmatrix} U^T,$$

where $\tilde{V}$ is an orthogonal matrix. Hence the singular values of H are the square roots of the eigenvalues of

$$\begin{pmatrix} I_i & O \\ O & X \end{pmatrix}, \quad \text{with} \quad X = \begin{pmatrix} C^2 & O \\ O & O \end{pmatrix} + \begin{pmatrix} S & O \\ O & I \end{pmatrix} G_s \begin{pmatrix} S & O \\ O & I \end{pmatrix}$$

where $G_s = (g_2^T, g_3^T)^T (g_2^T, g_3^T)$. We have

$$\lambda_l(X) = 1 - \lambda_{j+k-l+1}\left(\begin{pmatrix} S & O \\ O & I \end{pmatrix} (I - G_s) \begin{pmatrix} S & O \\ O & I \end{pmatrix}\right)$$

and

$$
\lambda_l\left(\begin{pmatrix} S & O \\ O & I \end{pmatrix} (I - G_s) \begin{pmatrix} S & O \\ O & I \end{pmatrix}\right) = \sigma_l^2\left((I - G_s)^{\frac{1}{2}} \begin{pmatrix} S & O \\ O & I \end{pmatrix}\right)
$$

$$(5.2)$$

$$
\geq \sigma_{\min}^2\left((I - G_s)^{\frac{1}{2}}\right) \sigma_l^2\left(\begin{pmatrix} S & O \\ O & I \end{pmatrix}\right).
$$

Since the diagonal elements of S are given in non-decreasing order and those of C are in non-increasing order, and

$$\sigma_{\min}^2\left((I - G_s)^{\frac{1}{2}}\right) = 1 - (\|g_2\|^2 + \|g_2\|^2) = \eta^2,$$

we have

$$\lambda_{j+k-l+1}\left(\begin{pmatrix} S & O \\ O & I \end{pmatrix} (I - G_s) \begin{pmatrix} S & O \\ O & I \end{pmatrix}\right) \geq \eta^2 \sigma_l^2\left(\begin{pmatrix} S & O \\ O & I \end{pmatrix}\right).$$

Hence from $\sigma_l(A, B) = \alpha_l = \cos(\theta_l)$, it follows that

$$\sigma_l(A, [B, b]) \leq \sqrt{1 - \beta_l^2(1 - (1 - \eta^2))}$$

(5.3)
$$= \sqrt{\alpha_l^2 + (1 - \alpha_l^2)(1 - \eta^2)}$$

$$= \sigma_l(A, B)\sqrt{1 + (1 - \eta^2)\tan^2\theta_l}.$$

Now we have that X is a rank one update of $\mathrm{diag}(C^2, O)$. Hence X has at most one additional nonzero eigenvalue. $\qquad\Box$

Here is how the numerical computation of the updating proceeds. Suppose Householder transformations or Jacobi rotations are utilized to compute the orthonormal basis of A and B so that we have [7, Section 5.2.]

$$B = Q\begin{pmatrix} R_B \\ O \end{pmatrix} = (Q_B, Q_B^\perp)\begin{pmatrix} R_B \\ O \end{pmatrix}.$$

Let $Q^T b = (x^T, \hat{b}^T)^T$. Apply a Householder transformation or a sequence of Jacobi rotations Q_b, we get

$$Q_b^T \hat{b} = \begin{pmatrix} \|\hat{b}\|_2 \\ 0 \end{pmatrix}, \quad \hat{b} = Q_b\begin{pmatrix} \|\hat{b}\|_2 \\ 0 \end{pmatrix}.$$

The QR decomposition of $[B, b]$ can be written as

$$[B, b] = [Q_B, Q_B^\perp Q_b]\begin{pmatrix} R_B & x \\ O & \|\hat{b}\|_2 \\ O & O \end{pmatrix}.$$

Then $[Q_B, Q_B^\perp(\hat{b}/\|\hat{b}\|_2)]$ is the orthonormal basis of $[B, b]$.

Remark 5.1. We note that the modified Gram-Schmidt algorithm could also be used to compute the orthonormal basis of A and B. Although it is twice as fast as the Householder transformation based algorithms, it has the drawback that the orthonormality of the computed basis depends on the condition numbers $\kappa_2(A)$ and $\kappa_2(B)$.[6]

For updating the SVD in (5.1), both the secular equation method (or the bisection method) [7, Section 8.6.3.], or the two-way chasing method [21] can be used. For a more detailed account of the bisection method the reader is referred to [5].

5.3. Large sparse or structured matrices. If the matrix pair (A, B) is large sparse or structured, then explicit computation of the orthonormal basis of A and B will usually gives rise to a dense matrix or destroys the

---

[6] It is our belief that $\kappa_2(A)$ and $\kappa_2(B)$ in the error analysis by Björck [3] can be replaced by the condition numbers defined in Section 4.

underlying structure. The purpose of this subsection is to propose a class of algorithms that will avoid explicit formation of the orthonormal basis of span$\{A\}$ and span$\{B\}$. Let us first consider a simple case: let A consist of one column, say a. Also, let the orthogonal projection onto span$\{B\}$ be P_B. Then the canonical correlation of the matrix pair (a, B) is given by

$$\sigma(a, B) = |(P_B a)^T a| / \|P_B a\|_2 \|a\|_2 \,,$$

and the canonical vectors are $a/\|a\|_2$ and $P_B a/\|P_B a\|_2$. Since $P_B a$ can be obtained by solving the following least squares problem:

$$\min_{x \in R^q} \|a - Bx\|_2 := \|a - Bx_0\|_2, \quad P_B a = Bx_0 \,,$$

the sparsity or structure of the matrix B can be fully exploited. For example, if the LSQR algorithm (cf. [13])is used to solve the above least squares problem, the matrix B is only used to form the matrix products Bx and $B^T y$ for given vectors x and y.[7]

In the general case, we propose the following alternating least squares (ALS) method to compute the largest canonical correlation of the matrix pair (A, B).

Algorithm 5.2. Choose $b_0 \in span\{B\}$ with $\|b_0\|_2 = 1$.

For $k = 0, 1, 2, \cdots$ until convergence do

1. Solve linear least squares problem:

$$\min_{x \in R^q} \|b_k - Ax\|_2 = \|b_k - Ax_k\|_2, \quad and\ form\ \ a_k = Ax_k/\|Ax_k\|_2\,;$$

2. Solve linear least squares problem:

$$\min_{y \in R^q} \|a_k - By\|_2 = \|a_k - By_k\|_2, \quad and\ form\ \ b_{k+1} = By_k/\|By_k\|_2\,;$$

Iterate.

Assume convergence in K steps. Now we compute

$$\sigma_1(A, B) = |b_K^T a_K|, \quad u_1 = a_K/\|a_K\|_2, \quad v_1 = b_K/\|b_K\|_2\,.$$

For the convergence criterion, we choose either

$$\|b_{k+1}^T a_{k+1}| - |b_k^T a_k\|, \quad or\ \min\{\|a_{k+1} - a_k\|_2, \|b_{k+1} - b_k\|_2\}$$

be below a certain given tolerance.

Remark 5.2. The alternating least squares method is an old and natural idea, which goes back to J. Von Neumann. It has been used extensively in the psychometrics literature, and a recent application can be found in [17].

---

[7] For a detailed presentation of fast algorithms for computing a matrix-vector product with Hankel or Toeplitz matrices, the reader is referred to [19].

Convergence analysis of the ALS method. We relate the ALS algorithm to a variant of the power method, and thus derive its convergence rate. First let us consider the power method. Since finding the canonical correlations is equivalent to computing the SVD of $Q_B^T Q_A$. Let

$$(5.4) \qquad T = \begin{pmatrix} O & (Q_B^T Q_A)^T \\ Q_B^T Q_A & O \end{pmatrix},$$

then the eigenvalues of T are $\{\pm \sigma_i(Q_B^T Q_A)\}$. Applying the power method to T, we have

$$(5.5) \qquad z_{k+1} = T z_k, \text{ with } z_0 \text{ an initital vector}.$$

Let $z_k = (x_k^T, y_k^T)^T$; equation (5.5) can be written as

$$x_{k+1} = Q_A^T Q_B y_k, \quad y_{k+1} = Q_B^T Q_A x_k.$$

We can use the most recent x_{k+1} to compute y_{k+1} so that

$$x_{k+1} = Q_A^T Q_B y_k, \quad y_{k+1} = Q_B^T Q_A x_{k+1}.$$

It follows that

$$Q_A x_{k+1} = Q_A Q_A^T Q_B y_k, \quad Q_B y_{k+1} = Q_B Q_B^T Q_A x_{k+1}.$$

The above two equations are equivalent, since we assume that A and B are of full column rank. Define

$$\tilde{x}_k = Q_A x_k, \quad \tilde{y}_k = Q_B y_k,$$

we have the following modified power method.

Algorithm 5.3. Choose $y_0 \in span\{B\}$ with $\|y_0\|_2 = 1$.

For $k = 0, 1, 2, \cdots$ until convergence do

$$x_{k+1} = Q_A Q_A^T \tilde{y}_k, \quad \tilde{x}_{k+1} = x_{k+1} / \|x_{k+1}\|_2 ;$$

$$y_{k+1} = Q_B Q_B^T \tilde{x}_{k+1} \quad \tilde{y}_{k+1} = y_{k+1} / \|y_{k+1}\|_2 ;$$

Iterate.

To see that Algorithm 5.3 is equivalent to the ALS algorithm, we observe that $b_0 \in \mathcal{R}(B)$, and can be written as $b_0 = Q_B s$ for some vector s. The solution of the least squares problem

$$\min_{x \in R^q} \|b_0 - Ax\|_2 = \|b_0 - Ax_0\|_2$$

is given by $x_0 = A^\dagger b_0$. Hence $a_0 = \gamma_0 Q_A Q_A^T (Q_B s)$, where γ_0 is the normalization factor. By induction we can prove

$$a_k = \gamma_k Q_A [(Q_B^T Q_A)^T (Q_B^T Q_A)]^k Q_A^T Q_B s; \quad b_k = \delta_k Q_B [(Q_B^T Q_A)(Q_B^T Q_A)^T]^k s.$$

where γ_k and δ_k are the normalization factors. Therefore the convergence rate of the ALS algorithm is dependent on

$$\kappa = (\sigma_2(Q_B^T Q_A)/\sigma_1(Q_B^T Q_A))^2 = (\sigma_2(A, B)/\sigma_1(A, B))^2 .$$

Example 5.1. We consider the matrix pair

$$A = U \begin{pmatrix} 1 & 0 \\ 0 & .8 \\ 0 & .6 \end{pmatrix} P_1 , \quad B = U \begin{pmatrix} 1 & 0 \\ 0 & 1 \\ 0 & 0 \end{pmatrix} P_2,$$

where U is an orthogonal matrix and P_1 and P_2 are nonsingular matrices. The canonical correlations are $\sigma_2(A, B) = 1, \sigma_2(A, B) = 0.8$. Therefore the convergence rate of the ALS algorithm is 0.64. We compute $\log_e(0.64)$

$$\log_e(0.64) = -.44628710262842 .$$

We have truncated the data at both ends. The best computed linear polynomial fit to the computed data gives the slope

$$-.44614014758065 ,$$

which matches the convergence rate quite well.

In the literature, various ways to accelerate the power method are given [7, Chapter 10]. We can adapt these acceleration schemes to the ALS algorithm, but we will not go into the details here.

The drawback of the ALS algorithm together with its various acceleration schemes is that only the largest canonical correlation is computed. To compute several canonical correlations at the same time, we can use certain versions of subspace iteration and we use various acceleration schemes. We will not discuss these extensions here, but instead we will show how to adapt the Lanczos method to our problem by using a similar idea as the ALS method. We apply the Lanczos algorithm [7, Chapter 9] and start with the matrix T defined in (5.4).

Algorithm 5.4. (*Lanczos Algorithm*)

$$
\begin{aligned}
&v_1 \text{ is given with } \|v_1\|_2 = 1 \\
&p_0 = v_1, \beta_0 = 1, \quad j = 0, u_0 = 0 \\
&\textbf{while } \beta_j \neq 0 \\
&\quad v_{j+1} = p_j/\beta_j; j = j + 1 \\
&\quad r_j = Q_B^T Q_A v_j - \beta_{j-1} u_{j-1} \\
&\quad \alpha_j = \|r_j\|_2; \quad u_j = r_j/\alpha_j \\
&\quad p_j = Q_A^T Q_B u_j - \alpha_j v_j \\
&\quad \beta_j = \|p_j\|_2 \\
&\textbf{end}
\end{aligned}
$$

We observe that the operator $Q_B^T Q_A$ is not available, since we do not explicitly form the orthonormal bases for span$\{A\}$ and span$\{B\}$. The device we use is to make a bases transformation. Let us transform the vectors generated in Algorithm 5.4 to the column spaces of A and B, i.e., span$\{A\}$ and span$\{B\}$, and denote

$$\tilde{u}_j = Q_B u_j, \ \tilde{v}_j = Q_A v_j, \ \tilde{r}_j = Q_B r_j, \ \tilde{p}_j = Q_A p_j.$$

Rewrite Algorithm 5.4 in the new basis, we obtain the following algorithm

Algorithm 5.5. (*Modified Lanczos Algorithm*)

$$
\begin{aligned}
&Choose\ v = As;\ set \tilde{v}_1 = v/\|v\|_2\\
&\tilde{p}_0 = v_1, \beta_0 = 1, j = 0, \tilde{u}_0 = 0\\
&\textbf{while } \beta_j \neq 0\\
&\qquad \tilde{v}_{j+1} = \tilde{p}_j/\beta_j; \quad j = j+1\\
&\qquad \tilde{r}_j = Q_B Q_B^T \tilde{v}_j - \beta_{j-1}\tilde{u}_{j-1}\\
&\qquad \alpha_j = \|\tilde{r}_j\|_2; \quad \tilde{u}_j = \tilde{r}_j/\alpha_j\\
&\qquad \tilde{p}_j = Q_A Q_A^T \tilde{u}_j - \alpha_j \tilde{v}_j\\
&\qquad \beta_j = \|\tilde{p}_j\|_2\\
&\textbf{end}
\end{aligned}
$$

Note that β_j and α_j in Algorithm 5.5. is the same as those in Algorithm 5.4. The computation of $Q_B Q_B^T \tilde{v}_j$ and $Q_A Q_A^T \tilde{u}_j$ are again carried out by solving the least squares problems:

$$(5.6) \qquad \min_{y \in R^q} \|\tilde{v}_j - By\|_2 = \|\tilde{v}_j - By_j\|_2 .$$

Then $Q_B Q_B^T \tilde{v}_j = By_j$. Similarly,

$$(5.7) \qquad \min_{x \in R^p} \|\tilde{u}_j - Ax\|_2 = \|\tilde{u}_j - Ax_j\|_2 ,$$

and $Q_A Q_A^T \tilde{u}_j = Ax_j$.

Remark 5.3. The above algorithm can be easily adapted to computing the canonical correlations between two linear subspaces defined either by the range space or null space of matrices. If for example, one of the subspace is defined by the null space of A, then instead of using $Q_A Q_A^T$ in the above, we use $I - Q_A Q_A^T$.

We have also tested the *Modified Lanczos Algorithm*. The matrix pair is given as follows

$$A = U \begin{pmatrix} C \\ S \end{pmatrix} P_1, \quad B = U \begin{pmatrix} I_n \\ O \end{pmatrix} P_2 ,$$

where U is orthogonal and P_1 and P_2 are nonsingular, with

$$C = \mathrm{diag}(0, 1/n, 2/n, \cdots, (n-1)/n), \quad \text{and} \quad S = \sqrt{(I_n - C^2)} .$$

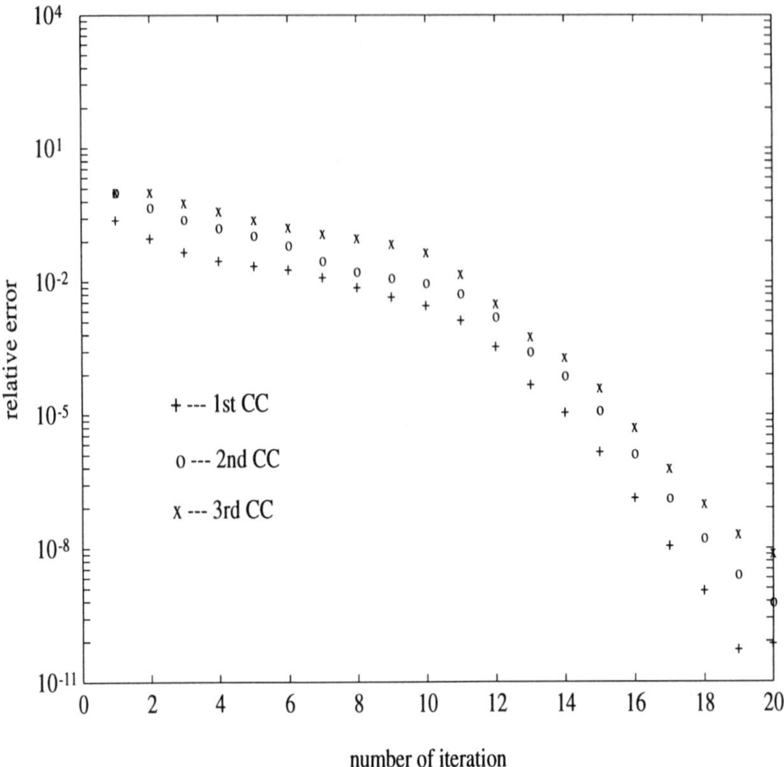

FIG. 5.1. *Convergence behavior of the modified Lanczos method*

Therefore the canonical correlations of (A, B) are $0, 1/n, 2/n, \cdots, (n-1)/n$. For the particular example in Figure 5.1, we chose $n = 100$. We do not solve the least squares problems in (5.6) and (5.7) exactly, instead we simulate the LSQR algorithm [13] by first using a direct method to solve the least squares problem and add noise to the solution. More numerical experiments using the LSQR will be carried out in the future. In Figure 5.1, the relative errors of the first three computed canonical correlations are plotted against the iteration numbers.

There remain a number of problems associated with this technique such as determining a preconditioner for solving the least squares problem. Nevertheless, we feel that the approach is of great potential use in computing canonical correlations of large or sparse matrix pairs and it certainly deserves further investigation.

REFERENCES

[1] T.W.ANDERSON, *An introduction to multivariate statistical analysis*, John Wiley and Sons, New York 1958.

[2] P.BESSE, *Edudie descriptive d'un process*, Ph.D. thesis, Paul-Sabatier University 1979.

[3] A.BJÖRCK, *Solving linear least squares problems by Gram-Schmidt orthogonalization*, BIT **7** (1967), 1–21.

[4] A.BJÖRCK, G.H.GOLUB, *Numerical methods for computing angles between linear subspaces*, Mathematics of Computation **27** (1973), 579–594.

[5] J.DEMMEL, W.GRAGG, *On computing accurate singular values and eigenvalues of acyclic matrices*, IMA Preprint Series **962**, IMA, University of Minnesota 1992.

[6] Y.ESCOUFIER, *Operators related to a data matrix*, in Recent Developments in Statistics (Amsterdam), North Holland 1976.

[7] G.H.GOLUB, C.F.VAN LOAN, *Matrix computations*, 2nd ed, Johns Hopkins University Press, Baltimore, Maryland 1989.

[8] H.HOTELLING, *Relation between two sets of variates*, Biometrika **28** (1936), 322–377.

[9] A.ISRAËLS, *Eigenvalue techniques for qualitative data*, DSWO Press, Leiden 1987.

[10] C.JORDAN, *Essai sur la géométrie à n dimensions*, Bulletin de la Société Mathématique **3** (1875), 103–174.

[11] J.R.KETTENRING, *Canonical analysis of several sets of variates*, Biometrika **58** (1971), 433–451.

[12] C.C.PAIGE, *A note on a result of Sun Ji-guang: sensitivity of the cs and gsv decomposition*, SIAM Journal on Numerical Analysis **21** (1984), 186–191.

[13] C.C.PAIGE, M.A.SAUNDERS, *LSQR: an algorithm for sparse linear equations and sparse least squares*, ACM Transaction on Mathematical Software **8** (1982), 43–71.

[14] C.R.RAO, H.YANAI, *General definition and decomposition of projectors and some application to statistical problems*, J. Statistical Planning and Inference **3** (1979), 1–17.

[15] G.W.STEWART, (Remarks made at an IMA workshop), 1992.

[16] G.W.STEWART, G.-J.SUN, *Matrix perturbation theory*, Academic Press, Boston 1990.

[17] E.VAN DER BURG, *Nonlinear canonical correlation and some related technique*, DSWO Press, Leiden 1988.

[18] P.-Å.WEDIN, *On angles between subspaces*, Matrix Pencils, B.Kågström, A.Ruhe, (eds.) Springer, New York 1983, 263–285.

[19] G.XU, T.KAILATH, *Fast signal-subspace decompsotion—part I: ideal covariance matrices*, (Manuscript submitted to ASSP), Information Systems Laboratory, Stanford University 1990.

[20] H.ZHA, *The restricted singular value decomposition of matrix triplets*, SIAM Journal on Matrix Analysis and Applications **12** (1991), 172–194.

[21] H.ZHA, *A two-way chasing scheme for reducing a symmetric arrowhead matrix to tridiagonal form*, Numerical Linear Algebra with Applications **1** (1992), 49–57.

CONTINUITY OF THE JOINT SPECTRAL RADIUS: APPLICATION TO WAVELETS

CHRISTOPHER HEIL* AND GILBERT STRANG†

Abstract. The joint spectral radius is the extension to two or more matrices of the (ordinary) spectral radius $\rho(A) = \max |\lambda_i(A)| = \lim \|A^m\|^{1/m}$. The extension allows matrix products Π_m taken in all orders, so that norms and eigenvalues are difficult to estimate. We show that the limiting process does yield a continuous function of the original matrices—this is their joint spectral radius. Then we describe the construction of wavelets from a dilation equation with coefficients c_k. We connect the continuity of those wavelets to the value of the joint spectral radius of two matrices whose entries are formed from the c_k.

1. Introduction. The (ordinary) *spectral radius* of a matrix A is the magnitude of its largest eigenvalue:

$$\rho = \rho(A) = \max\{|\lambda| : \lambda \text{ is an eigenvalue of } A\}.$$

This number controls the growth or decay of the powers A^m. If $\rho < 1$ then $A^m \to 0$ as $m \to \infty$. If $\rho > 1$ then the matrix powers are unbounded. The marginal case $\rho = 1$ leaves boundedness undecided. In any norm the mth root of $\|A^m\|$ always converges to ρ as $m \to \infty$.

We describe below how the degree of continuity of a wavelet (expressed by its Hölder exponent α) is controlled by a spectral radius. But there is a crucial difference from $\rho(A)$: *two matrices are involved instead of one*. The number $\rho(A)$ becomes a *joint spectral radius* $\hat{\rho}(A, B)$. It is still defined by a limit of mth roots, but $\|A^m\|$ is replaced by the largest norm $\|\Pi_m\|$ of products of A's and B's. The product $\Pi_m = ABAAB \cdots$ has its m factors in any order:

$$(1.1) \qquad \hat{\rho} = \hat{\rho}(A, B) = \lim_{m \to \infty} \left(\max \|\Pi_m\|\right)^{1/m}.$$

For symmetric or normal or commuting or upper-triangular matrices, this joint spectral radius is the larger of $\rho(A)$ and $\rho(B)$. Always $\hat{\rho}(A, B) \geq \rho(A)$ and $\hat{\rho}(A, B) \geq \rho(B)$, since the product Π_m might be A^m or B^m. An extreme case of inequality is

$$A = \begin{pmatrix} 0 & 2 \\ 0 & 0 \end{pmatrix} \quad \text{and} \quad B = \begin{pmatrix} 0 & 0 \\ 2 & 0 \end{pmatrix} \quad \text{and} \quad AB = \begin{pmatrix} 4 & 0 \\ 0 & 0 \end{pmatrix}.$$

The eigenvalues of A and B are all zero, so that $\rho(A) = 0 = \rho(B)$. In fact A^2 and B^2 are zero matrices, so the product Π_m is nonzero only when

---

* School of Mathematics, Skiles Building, Room 242, Georgia Institute of Technology, Atlanta, Georgia 30332.

† Department of Mathematics, Massachusetts Institute of Technology, Cambridge, Massachusetts 02139. Partially supported by National Science Foundation Grant DMS-9006220.

factors A and B alternate. The key to $\hat{\rho}$ is that $\rho(AB)^{1/2} = 2 = \|AB\|^{1/2}$. Therefore $\hat{\rho}(A, B)$, which is between eigenvalues and norms, also equals 2.

The eigenvalues of A and B fail to control the eigenvalues of products— not to mention the norms of those products. We cannot compute $\hat{\rho}$ from $\rho(A)$ and $\rho(B)$.

The spectral radius is bounded above by norms and below by eigenvalues. For a single matrix we have (for each m) an equality and an inequality:

$$\rho(A^m)^{1/m} = \rho(A) \leq \|A^m\|^{1/m}.$$

For two matrices we maximize over products in all orders and we expect two inequalities:

$$(1.2) \qquad (\max \rho(\Pi_m))^{1/m} \leq \hat{\rho}(A, B) \leq (\max \|\Pi_m\|)^{1/m}.$$

The proof of the first inequality comes from the corresponding result for a single matrix, by considering repetitions $\Pi_m \Pi_m \cdots \Pi_m$ of any fixed product. For the second inequality, break any product Π_n into pieces of fixed length m with a remainder: if $n = mq + r$ then $\Pi_n = \Pi_m^{(1)} \Pi_m^{(2)} \cdots \Pi_m^{(q)} \Pi_r$. For $n = mq$ and $r = 0$ we have $\|\Pi_n\|^{1/n} \leq (\max \|\Pi_m\|)^{1/m}$. The extra factor Π_r has no effect in the limit as $n \to \infty$.

The right side of (1.2) approaches equality as $m \to \infty$ by the definition of the joint spectral radius. Whether the left side also approaches equality is much less clear. It is a beautiful theorem of Berger and Wang [BW1] that this does occur:

THEOREM 1.1. [BW1]. $\displaystyle\limsup_{m \to \infty} (\max \rho(\Pi_m))^{1/m} = \hat{\rho}(A, B)$.

Thus $\hat{\rho}$ can be approximated from above and from below, by computing the norms and eigenvalues of finitely many matrix products. The convergence as $m \to \infty$ may be quite slow. Examples are given by Colella and Heil [CH1], [HC], with a recursive algorithm that significantly reduces the calculation on the norm side.

EXAMPLE 1.1. Set

$$A = \tfrac{1}{5} \begin{pmatrix} 3 & 0 \\ 1 & 3 \end{pmatrix} \qquad \text{and} \qquad B = \tfrac{1}{5} \begin{pmatrix} 3 & -3 \\ 0 & -1 \end{pmatrix}.$$

Then by actual computations up to $m = 30$ factors,

$$\hat{\rho} \geq \max_{1 \leq m \leq 30} (\max \rho(\Pi_m))^{1/m} = \rho(A^{12}B)^{1/13} \approx 0.659679$$

and

$$\hat{\rho} \leq \min_{1 \leq m \leq 30} (\max \|\Pi_m\|)^{1/m} = \|A^5 B A^{11} B A^{12}\|^{1/30} \approx 0.671271.$$

Less than two decimal places of accuracy are achieved after computing $2^{31} - 2$ matrix products. With 612944 products in the recursive algorithm the estimate is $\hat{\rho} \leq 0.660025$.

It is frustrating that this fundamental number $\hat{\rho}$, a function only of the entries in A and B, should be so difficult to compute. The same is true in the more highly developed theory of products of *random matrices* [CKN]. The corresponding number δ is an expected value instead of a maximum, and the only weakness is its resistance to calculation. In our present problem, Lagarias and Wang [LW] conjecture that equality holds on the left side of (1.2) for some finite m. Thus an eigenvalue of a finite product Π_m (for unknown and possibly large m) may reveal the exact value of the joint spectral radius.

Little is known about $\hat{\rho}$. From its definition (1.1), we were not even certain that $\hat{\rho}$ was a continuous function. Certainly each norm $\|\Pi_m\|^{1/m}$ depends continuously on the entries of A and B. This assures us that the limit $\hat{\rho}$ is at least upper-semicontinuous. (It is an infimum of continuous functions, the norms of products.) The Berger–Wang theorem yields the opposite result, that $\hat{\rho}$ is at the same time lower-semicontinuous. (It is a supremum of continuous functions, the eigenvalues of products.) Eventually we realized, with this essential help from Berger and Wang, that continuity does hold. This is the unique novel result of the present note:

COROLLARY 1.1. *$\hat{\rho}(A, B)$ is a continuous function of the matrices A and B.*

Allow us to write out the proof in what may be unnecessary detail. We are given matrices A and B and a fixed $\varepsilon > 0$. For sufficiently large m, both inequalities in (1.2) are within $\varepsilon/2$ of equality. Then if $\|A - C\|$ and $\|B - D\|$ are small enough, each side of (1.2) is within $\varepsilon/2$ of the corresponding inequality for C and D. Therefore $\hat{\rho}(C, D)$, which is caught in between, is within ε of $\hat{\rho}(A, B)$.

2. Wavelets and linear algebra. A wavelet is a function ψ whose dilates and translates $\psi(2^j x - k)$ form an orthogonal basis for $L^2(R)$. (More general definitions are possible. There are biorthogonal wavelets and even nonorthogonal wavelets.) We briefly describe the classical construction. The discrete version yields a particularly attractive orthonormal basis for R^n. The expansion of a vector in terms of this discrete basis is achieved by a very fast algorithm—the discrete wavelet transform operates in $\mathcal{O}(n)$ steps while the discrete Fourier transform requires $\mathcal{O}(n \log n)$. The expository paper [S] describes analogies and competitions between these transforms. In most of signal processing the standard methods are Fourier-based. In the compression of fingerprint images (the FBI has 25 million to digitize and compare) wavelet bases now seem to be superior.

The joint spectral radius enters in determining the smoothness (the order of Hölder continuity) of wavelets. The matrices A and B contain coefficients from the *dilation equation*. This has become the starting point for the construction of ψ, and we briefly outline the steps. The next section explains the connection to the joint spectral radius.

First, choose coefficients $(c_0, \ldots, c_N)$. Second, solve the dilation equation for the *scaling function* φ:

$$(2.1) \qquad \varphi(x) \;=\; \sum_{k=0}^{N} c_k\, \varphi(2x - k).$$

Third, construct ψ directly from φ and the c_k (in reverse order and with alternating signs):

$$(2.2) \qquad \psi(x) \;=\; \sum_{k=0}^{N} (-1)^k\, c_{N-k}\, \varphi(2x - k).$$

(Other choices of indexing are possible; this gives φ and ψ supported in $[0, N]$). The properties of ψ clearly depend on the choice of the c_k. The condition for "minimal accuracy" is

$$(2.3) \qquad \sum_{k} c_{2k} \;=\; \sum_{k} c_{2k+1} \;=\; 1.$$

The condition for orthogonality of the family $\psi(2^j x - k)$ is

$$(2.4) \qquad \sum_{k} c_k\, \bar{c}_{k+2j} \;=\; 2\, \delta_{0j}.$$

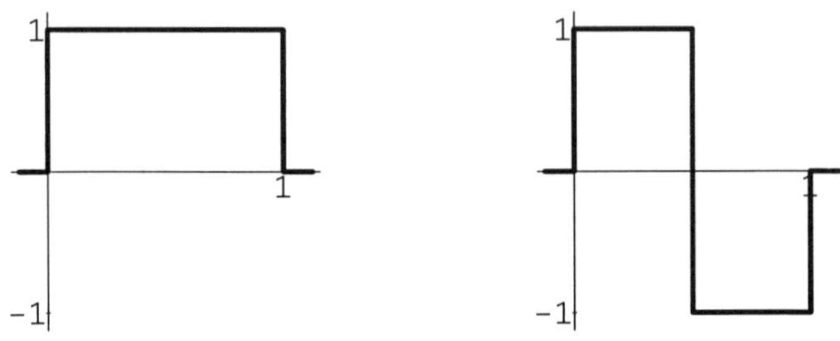

FIG. 1. *Box function φ (left) and Haar wavelet ψ (right).*

EXAMPLE 2.1. Make the choice $c_0 = c_1 = 1$. Then the dilation equation (2.1) requires $\varphi(x)$ to be the sum of $\varphi(2x)$ and $\varphi(2x - 1)$. The solution is the box function $\varphi = \chi_{[0,1)}$. Both the minimal accuracy and orthogonality conditions hold, and $\psi(x) = \varphi(2x) - \varphi(2x - 1)$ generates a wavelet orthonormal basis. This property of ψ was first observed by Haar [H].

The accuracy and orthogonality conditions (2.3) and (2.4) ensure that equation (2.1) yields a unique, compactly supported, integrable scaling function φ. Without these conditions, existence is not guaranteed. (If a compactly supported scaling function does exist then it is unique.) We only consider compact support in this note.

3. The matrix form of the construction. To make the connection with the joint spectral radius, we convert the dilation equation (2.1) into a matrix form. The key fact is that if a compactly supported scaling function φ exists then it is supported in the interval $[0, N]$. Therefore, the vector-valued function v defined by

$$v(x) \;=\; \begin{pmatrix} \varphi(x) \\ \varphi(x+1) \\ \vdots \\ \varphi(x+N-1) \end{pmatrix}, \qquad \text{for } x \in [0,1],$$

"captures" all the essential information about φ. Assume φ is continuous, so that $\varphi(0) = 0 = \varphi(N)$. Choose x in the interval $[0, 1/2]$, and apply the dilation equation to each of $\varphi(x), \varphi(x+1), \ldots, \varphi(x+N-1)$. Working through some algebra, we find that these values of φ are determined by the values of $\varphi(2x), \varphi(2x+1), \ldots, \varphi(2x+N-1)$ in a fixed linear manner. In other words, there is a linear transformation T_0 which takes $v(2x)$ to $v(x)$ for each $0 \leq x \leq 1/2$:

$$(3.1) \qquad\qquad v(x) \;=\; T_0 v(2x), \qquad \text{for } 0 \leq x \leq 1/2.$$

Similarly, there is a T_1 such that

$$(3.2) \qquad\qquad v(x) \;=\; T_1 v(2x-1), \qquad \text{for } 1/2 \leq x \leq 1.$$

In fact, T_0 and T_1 are the $N \times N$ matrices whose entries are $(T_0)_{ij} = c_{2i-j-1}$ and $(T_1)_{ij} = c_{2i-j}$. There is *consistency* at $x = 1/2$:

$$(3.3) \qquad\qquad v(1/2) \;=\; T_0 v(1) \;=\; T_1 v(0).$$

To simplify the notation, let τx be the fractional part of $2x$:

$$\tau x \;=\; (2x) \bmod 1 \;=\; \begin{cases} 2x, & 0 \leq x < 1/2, \\ 2x - 1, & 1/2 < x \leq 1. \end{cases}$$

We purposely leave $\tau(1/2)$ undefined. If $x = k/2^j$ then it has two possible binary expansions: a "terminating" expansion ending in infinitely many zeros, and another expansion ending in infinitely many ones. We call such points *dyadic*. Otherwise, x has a unique binary expansion. In any case, the first digit d_1 in the binary expansion $x = .d_1 d_2 \cdots$ is unique as long as $x \neq 1/2$:

$$d_1 = \begin{cases} 0, & 0 \le x < 1/2, \\ 1, & 1/2 < x \le 1 \end{cases}, \qquad \text{and} \qquad \tau x = .d_2 d_3 \cdots.$$

Therefore, except for $x = 1/2$, we can summarize (3.1) and (3.2) as a single equation:

$$(3.4) \qquad v(x) = T_{d_1} v(\tau x), \qquad \text{for } x = .d_1 d_2 \cdots.$$

Because of the consistency (3.3), this formula also applies to $x = 1/2$: either of the two binary expansions $1/2 = .100 \cdots$ or $1/2 = .011 \cdots$ may be used as long as T_{d_1} and $\tau(1/2)$ are interpreted consistently.

Now we use the assumption that φ is continuous. Let $x < y$ be dyadic points in $[0,1]$. If y is close enough to x then the first few digits in its terminating binary expansion will coincide with the first few digits in the corresponding expansion of x. That is, $x = .d_1 \cdots d_m d_{m+1} d_{m+2} \cdots$ and $y = .d_1 \cdots d_m d'_{m+1} d'_{m+2} \cdots$ for some m. The closer y is to x, the larger m will be. Applying (3.4) repeatedly, we obtain

$$(3.5) \qquad \begin{aligned} v(y) - v(x) &= T_{d_1}\big(v(\tau y) - v(\tau x)\big) \\ &= T_{d_1} T_{d_2}\big(v(\tau^2 y) - v(\tau^2 x)\big) \\ &\ \ \vdots \\ &= T_{d_1} \cdots T_{d_m}\big(v(\tau^m y) - v(\tau^m x)\big). \end{aligned}$$

Since $m \to \infty$ and $v(y) \to v(x)$ as $y \to x$, the products $\Pi_m = T_{d_1} \cdots T_{d_m}$ must converge to zero in the limit, at least when applied to vectors of the form $v(w) - v(z)$. To ensure that only vectors of this form are considered, we restrict our attention to the subspace

$$W = \text{span}\{v(w) - v(z) : \text{dyadic } w, z \in [0,1]\},$$

which is invariant under both T_0 and T_1. Then all the restricted products $(\Pi_m)|_W$ must converge to zero as $m \to \infty$. As Berger and Wang [BW1] observed, this is equivalent to $\hat{\rho} < 1$ (on W). We therefore have the following necessary condition for φ to be continuous:

THEOREM 3.1. [CH2] *If φ is a continuous scaling function then*

$$\hat{\rho}(T_0|_W, T_1|_W) < 1.$$

The subspace W is not as difficult to determine explicitly as it may appear. It is the smallest subspace which contains the vector $v(1) - v(0)$ and is invariant under both T_0 and T_1. By an appropriate change of basis we can always realize the action of T_0, T_1 on W as the action of two smaller matrices A, B on R^n with $n = \dim(W)$. In this case, $\hat{\rho}(T_0|_W, T_1|_W) = \hat{\rho}(A, B)$.

The arguments leading to Theorem 3.1 did not make use of the accuracy or orthogonality conditions (2.3) or (2.4). Theorem 3.1 therefore applies to completely arbitrary dilation equations. As an implicit corollary, we obtain a necessary condition for φ to be continuously differentiable. This follows since φ' is itself a scaling function for the dilation equation with coefficients doubled to $2c_k$:

$$\varphi'(x) = \sum_{k=0}^{N} 2\,c_k\,\varphi'(2x - k).$$

A necessary condition for $\varphi \in C^n$ follows by repetition. These corollaries are given explicitly in [HC].

When the minimal accuracy condition (2.3) does hold, the matrices T_0 and T_1 are column stochastic in the sense that each column sums to one—although the entries need not be nonnegative. Therefore the row vector $(1, \ldots, 1)$ is a common left eigenvector for T_0 and T_1 for the eigenvalue 1. The subspace

$$(3.6) \qquad\qquad V = \{u \in C^N : u_1 + \cdots + u_N = 0\},$$

is therefore invariant under both T_0 and T_1, and V contains W.

We can also work out a converse to Theorem 3.1 in terms of a joint spectral radius. This time we must first construct the vector-valued function v and from it obtain a scaling function φ. Our inspiration again comes from applying (3.4) recursively: if x is a dyadic point with terminating binary expansion $x = .d_1 \cdots d_m$ then $\tau^m x = 0$, so

$$
\begin{aligned}
v(x) \;&=\; T_{d_1} v(\tau x) \\
&\;\;\vdots \\
&=\; T_{d_1} \cdots T_{d_m} v(\tau^m x) \\
&=\; T_{d_1} \cdots T_{d_m} v(0) \\
&=\; \Pi_m v(0).
\end{aligned}
$$

$$(3.7)$$

Thus $v(0)$ determines $v(x)$ for each dyadic x. To find $v(0)$, solve the eigenvector problem $v(0) = T_0 v(0)$. Our recursion (3.7) amounts to the observation that once the values of φ are known at the integers, the dilation equation gives the values at the half-integers, then the quarter-integers, and so forth to all dyadic points. This is a fast recursive algorithm for graphing any scaling function φ.

Now that the values of $v(x)$ have been constructed for dyadic x, we can fill in the values at intermediate points by taking limits—if this v is continuous on the set of dyadic points. In this case we "unfold" v to get a continuous scaling function φ by defining

$$\varphi(x) = \begin{cases} 0, & x \le 0 \text{ or } x \ge N, \\ v_i(x), & i-1 \le x \le i, \ i = 1, \ldots, N. \end{cases}$$

Here $v_i(x)$ is the ith component of $v(x)$. So, the problem is to ensure continuity of v on the dyadics. The argument is similar to the one used for Theorem 3.1. If $x = .d_1 \cdots d_m d_{m+1} d_{m+2} \cdots$ and $y = .d_1 \cdots d_m d'_{m+1} d'_{m+2} \cdots$ with x, y both dyadic then (3.5) holds. If we assume that all the restricted products $(\Pi_m)|_W$ converge to zero as $m \to \infty$ then we will have $v(y) \to v(x)$ as $y \to x$. But the convergence of all products to zero is equivalent to the assumption $\hat{\rho} < 1$. So, we have the following sufficient condition for the existence of a continuous scaling function:

THEOREM 3.2. [CH2] *If $\hat{\rho}(T_0|_W, T_1|_W) < 1$ then the dilation equation yields a continuous scaling function φ.*

The actual details involved in the proofs of Theorems 3.1 and 3.2 reveal precise bounds for the possible Hölder exponents of φ and its derivatives. (The number $\alpha \leq 1$ is a Hölder exponent if there exists C_α so that $|\varphi(x) - \varphi(y)| \leq C_\alpha |x - y|^\alpha$ for all x, y.) In fact, φ is Hölder continuous for each exponent α in the range $0 \leq \alpha < -\log_2 \hat{\rho}$ and not for any $\alpha > -\log_2 \hat{\rho}$. (For our matrices T_0 and T_1, the joint spectral radius $\hat{\rho}$ is never below $1/2$ so that $-\log_2 \hat{\rho} \leq 1$.) A condition can also be given for the marginal case $\alpha = -\log_2 \hat{\rho}$. As with Theorem 3.1, Theorem 3.2 implicitly leads to a sufficient condition for $\varphi \in C^n$.

EXAMPLE 3.1. Make the choice $c_0 = 3/5$, $c_1 = 6/5$, $c_2 = 2/5$, $c_3 = -1/5$. The accuracy and orthogonality conditions (2.3) and (2.4) are met. The scaling function φ and wavelet ψ are shown in Figure 2. The subspace W equals the subspace V, and $\hat{\rho}(T_0|_W, T_1|_W) = \hat{\rho}(A, B)$ with A, B as in Example 1.1. Thus $\hat{\rho}(T_0|_W, T_1|_W) \approx 0.66 < 1$, so φ and ψ are continuous. The maximum Hölder exponent is $\alpha \approx -\log_2 0.66 \approx 0.60$. Since $\alpha < 1$, φ and ψ are not differentiable.

We close by mentioning some connections, observed in [CH2], between Corollary 1.1 and the continuity of the joint spectral radius. By Corollary 1.1, $\hat{\rho}(A, B)$ is a continuous function of the entries of A and B. The entries of T_0 and T_1 consist of the coefficients $(c_0, \ldots, c_N)$. Despite this, $\hat{\rho}(T_0|_W, T_1|_W)$ is not in general a continuous function of $(c_0, \ldots, c_N)$ because the dimension of W can change abruptly as the coefficients vary. However, if the minimal accuracy condition (2.3) holds then $W \subset V$, and V is *independent* of the coefficients. Therefore $\hat{\rho}(T_0|_V, T_1|_V)$ is continuous, and the condition $\hat{\rho}(T_0|_V, T_1|_V) < 1$ is stable under small perturbations of the coefficients. Then $\hat{\rho}(T_0|_W, T_1|_W) \leq \hat{\rho}(T_0|_V, T_1|_V) < 1$ ensures the existence of a continuous scaling function—which deforms uniformly as the coefficients vary. But the maximum Hölder exponents need not vary continuously since they depend critically on the value of $\hat{\rho}(T_0|_W, T_1|_W)$.

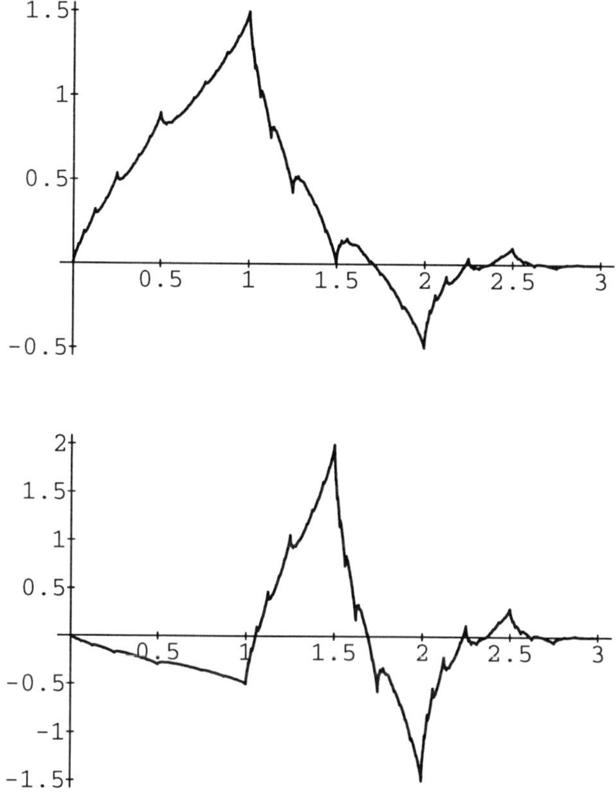

FIG. 2. *Scaling function φ (top) and wavelet ψ (bottom) from Example 3.1.*

4. Historical notes. The joint spectral radius was first described in [RS], for subsets of normed algebras. The first examples of continuous wavelets were found by Strömberg [St] and Meyer [Me]. The Meyer wavelet is C^∞, but not compactly supported. Mallat and Meyer later developed multiresolution analysis [Ma], which results in the wavelet construction outlined in Section 2. Daubechies [D] discovered the first examples of continuous, compactly supported wavelets, including C^n wavelets for arbitrarily large n. Later, with Lagarias, she proved that compactly supported wavelets cannot be infinitely differentiable [DL1].

Daubechies and Lagarias also derived the matrix method for estimating the smoothness of wavelets, on which our discussion in Section 3 is based. In [DL2], they derived the matrix form (7)–(8) of the dilation equation and proved that if the coefficients satisfy the minimal accuracy condition (2.3) and if $\hat{\rho}(T_0|_V, T_1|_V) < 1$ then φ is continuous with Hölder exponent at least $-\log_2 \hat{\rho}(T_0|_V, T_1|_V) - \varepsilon$. Theorem 3.2 is an extension of this result to arbitrary dilation equations, and Theorem 3.1 is its converse. These two theorems were proved by Colella and Heil [CH2] together with corresponding sharp bounds for the Hölder exponent. Berger and Wang [BW2] independently obtained the same results. Daubechies and Lagarias also derived sufficient conditions for n-times differentiability and used the matrix approach to study the *local* smoothness of scaling functions and wavelets [DL2].

Dilation equations play a key role in subdivision or refinement schemes used in computer aided graphics. This is a separate application, using different coefficients c_k and different functions φ, but there is substantial overlap with wavelet theory. In particular, Micchelli and Prautzsch [MP] published the matrix form of the dilation equation before Daubechies and Lagarias, and they proved a necessary and sufficient condition for continuity of a scaling function [MP]. This condition was not in terms of a joint spectral radius, and did not lead to the Hölder exponent. An excellent survey of subdivision schemes is [CDM].

REFERENCES

[BW1] M.A. Berger and Y. Wang, *Bounded semi-groups of matrices*, Lin. Alg. Appl., 166 (1992), pp. 21–27.

[BW2] M.A. Berger and Y. Wang, *Multi-scale dilation equations and iterated function systems*, Random Computational Dynamics (to appear).

[CDM] A. Cavaretta, W. Dahmen, and C.A. Micchelli, *Stationary Subdivision*, Mem. Amer. Math. Soc., 93 (1991), pp. 1–186.

[CH1] D. Colella and C. Heil, *The characterization of continuous, four coefficient scaling functions and wavelets*, IEEE Trans. Inf. Th., Special Issue on Wavelet Transforms and Multiresolution Signal Analysis, 38 (1992), pp. 876–881.

[CH2] D. Colella and C. Heil, *Characterizations of scaling functions: Continuous solutions*, SIAM J. Matrix Anal. Appl., 15 (1994), pp. 496–518.

[CKN] J.E. Cohen, H. Kesten, and C.M. Newman, eds., *Random Matrices and Their Applications*, Contemporary Math. 50, Amer. Math. Soc., Providence, 1986.

[D] I. Daubechies, *Orthonormal bases of compactly supported wavelets*, Comm. Pure Appl. Math., 41 (1988), pp. 909–996.

[DL1] I. Daubechies and J. Lagarias, *Two-scale difference equations: I. Existence and global regularity of solutions*, SIAM J. Math. Anal., 22 (1991), pp. 1388–1410.

[DL2] I. Daubechies and J. Lagarias, *Two-scale difference equations: II. Local regularity, infinite products of matrices and fractals*, SIAM J. Math. Anal., 23 (1992), pp. 1031–1079.

[H] A. Haar, *Zur Theorie der orthogonalen Funktionensysteme*, Math. Ann., 69 (1910), pp. 331–371.

[HC] C. Heil and D. Colella, *Dilation equations and the smoothness of compactly supported wavelets*, Wavelets: Mathematics and Applications, J. Benedetto and M. Frazier, eds., CRC Press (1993), pp. 161–200.

[LW] J.C. Lagarias and Y. Wang, *The finiteness conjecture for the generalized spectral radius of a set of matrices*, Lin. Alg. Appl. (to appear).

[Ma] S.G. Mallat, *Multiresolution approximations and wavelet orthonormal bases for $L^2(R)$*, Trans. Amer. Math. Soc., 315 (1989), pp. 69–87.

[Me] Y. Meyer, *Principe d'incertitude, bases hibertiennes et algèbres d'opérateurs*,

Séminaire Bourbaki, 662 (1985-1986).

[MP] C.A. MICCHELLI AND H. PRAUTZSCH, *Uniform refinement of curves*, Lin. Alg. Appl., 114/115 (1989), pp. 841–870.

[RS] G.C. ROTA AND G. STRANG, *A note on the joint spectral radius*, Kon. Nederl. Akad. Wet. Proc. A, 63 (1960), pp. 379–381.

[S] G. STRANG, *Wavelet transforms versus Fourier transforms*, Bull. Amer. Math. Soc. 28 (1993), pp. 288–305.

[St] J.O. STRÖMBERG, *A modified Franklin system and higher-order spline systems on R^n as unconditional bases for hardy spaces*, Conf. on Harm. Anal. in Honor of A. Zygmund, Vol. II, W. beckner et al., eds., Wadsworth (1981), pp. 475–494.

INVERSION OF GENERALIZED CAUCHY MATRICES AND OTHER CLASSES OF STRUCTURED MATRICES

GEORG HEINIG*

Abstract. Fast inversion algorithms for strongly nonsingular matrices of the form $C = \left[\frac{z_i^T y_j}{c_i - d_j}\right]$ (generalized Cauchy matrices), where z_i, y_j are column vectors and c_i, d_j are complex numbers, are presented. The approach is based on the interpretation of equations $C\xi = \eta$ as tangential interpolation problems. Furthermore, it is described how other types of structured matrices like Toeplitz matrices and their generalizations can be transformed into generalized Cauchy matrices. This transformation can be utilized in order to get stable algorithms.

Key words. structured matrix, Cauchy, Hilbert, Toeplitz matrix, fast algorithm, rational interpolation

AMS(MOS) subject classifications. 15A 09 (47B 35, 65F 20)

1. Introduction. In this paper we consider matrices of the form

$$(1.1) \qquad C = \left[\frac{z_i^T y_j}{c_i - d_j}\right]_{i,j=1}^{n},$$

where c_i, d_j are complex numbers such that $c_i \neq d_j$ for all i and j, and z_i, y_j are given column vectors from $\mathbf{C}^r$ and r is small compared with n. In the case $r = 1$ and $z_i = y_j = 1$ the matrix C is usually called *Cauchy matrix* or *generalized Hilbert matrix*. For this reason matrices 1.1 will be referred to as *generalized Cauchy* matrices. An important special class of generalized Cauchy matrices are Loewner matrices

$$\left[\frac{\xi_i - \eta_j}{c_i - d_j}\right],$$

which correspond to the special choice $r = 2$, $z_i = [\xi_i \quad - 1]^T$ and $y_j = [1 \quad \eta_j]^T$ and appear in rational interpolation (see [10]). Inversion formulas and algorithms for generalized Cauchy matrices were presented in [18], [13] and [14].

The present paper has two aims. First we want to show that fast inversion algorithms can easily be constructed if one utilizes the interpolation interpretation of equations of the form $C\xi = \eta$. Actually an equation of this form is equivalent to a tangential interpolation problem for rational vector functions. For the interpolation problem recursion formulas can be derived in a very natural way. Translating this into vector language

---

* Department of Mathematics, Kuwait University, P.O. Box 5969, Safat 13060 KUWAIT. email: georg@math-1.kuniv.edu.kw. The work was carried out during a visit at the IMA, University of Minnesota in Minneapolis, March 1992.

this leads to fast algorithms for solving systems $C\xi = \eta$ with complexity $O(n^2)$ or less.

The second aim of this paper is based on the following observation: A generalized Cauchy matrix remains to be a matrix of this type after any permutation of columns and rows. Other classes of structured matrices like Toeplitz and Hankel matrices and generalizations of them do not have this property. The advantage of the invariance of the class under permutations of rows and columns consists in the fact that it makes pivoting techniques possible. So one can construct algorithms which are not only fast but also stable.

To avoid the well-known stability problems occuring in inversion algorithms for indefinite or nonsymmetric Toeplitz and Hankel matrices (see [8]) we suggest to transform them into generalized Cauchy matrices. It will be shown that this can be done via discrete Fourier transformations and preserving the condition of the matrix. For the corresponding generalized Cauchy matrices c_i and d_j will be roots of unity. This fact also allows to construct new $O(n \log^2 n)$ complexity algorithms for Toeplitz and other structured matrices.

The algorithms for the inversion of generalized Cauchy matrices presented in the literature and also the algorithms presented below work only for strongly nonsingular matrices, i.e. for matrices $C = [c_{ij}]_1^n$ for which all principal submatrices $[c_{ij}]_1^k$ $(k = 1, \cdots, n)$ are nonsingular.

If one uses pivoting techniques this assumption is not really a restriction of generality. However, it seems to be also desirable to have algorithms working for arbitrary nonsingular matrices C without column and row permutation. The subsequent paper [16] will be deal with this problem. Let us remark that in order to do this one has to consider a wider class of matrices, namely generalized Cauchy-Vandermonde matrices. In this sense the present note can be regarded as preliminaries of the paper [16].

Let us shortly decribe the contents. In Section 2 we present an inversion formula involving the solution of certain "fundamental equations" which is a consequence of the fact that generalized Cauchy matrices satisfy Sylvester equations with a rank r right–hand side. The formula shows that the inverse of a generalized Cauchy matrix is such a matrix again. In Section 3 we give the interpolation interpretation of the fundamental equations, which leads to the concept of fundamental matrix. This concept was already used in the theory of Hankel and Toeplitz (see [17]) matrices and is related to the concept of resolvent matrix in classical interpolation problems (Nevanlinna–Pick, Schur, Nehari problems) and can be interpreted in the context of realization theory (cf. [2]). There are two fundamental matrices: a right and a left one. The realization interpretation shows that one is just the inverse of the other.

In Section 4 we present recursion formulas for the fundamental matrices. Translating this into vector language this gives recursions for the solutions of the fundamental equations by describing the connection between two adjoining nested submatrices of C. This leads to an $O(n^2)$ complexity algorithm to compute C^{-1}. The algorithm can also be used in order to evaluate the LU-factorizations of C^{-1} and C.

However, the algorithm described in Section 4 involves inner product calculations and is therefore not very convenient in parallel computation. The disadvantage can be avoided by computing recursively certain residuals. This results in a Schur-type algorithm which will be described in Section 5.

In Section 6 we show that the Schur-type algorithm can be accelerated using a divide-and-conquer strategy. The complexity obtained in this way is only $O(n \log^3 n)$ compared with $O(n \log^2 n)$ for Toeplitz matrices. In case that the c_i and d_j are unit roots the amount can be reduced to $O(n \log^2 n)$.

Sections 7 and 8 are dedicated to transformations of generalized Cauchy matrices. In Section 7 we show that many types of structured matrices can be transformed into generalized Cauchy matrices with c_i and d_j beeing unit roots. Special attention is paid to Toeplitz matrices and their generalizations, since in this case the transformation matrices are related to the unitary (and therefore condition preserving) Fourier matrices. In Section 8 we show generalized Cauchy matrices with arbitrary c_i and d_j can be transformed into close-to-Toeplitz matrices and to generalized Cauchy matrices with unit roots c_i and d_j. This leads to $O(n \log^2 n)$ complexity algorithms for generalized Cauchy systems.

2. Inversion formula. To begin with let us introduce some notations. For given $c = (c_i)_1^n \in \mathbf{C}^n$, let $D(c)$ denote the diagonal matrix

$$D(c) = \operatorname{diag}(c_1, \ldots, c_n)$$

Throughout the paper, let C denote a matrix given by 1.1. We introduce the matrices

$$Z = \operatorname{col}(z_i^T)_1^n, \quad Y = \operatorname{col}(y_j^T)_1^n .$$

and

$$V_k(c, Z) = [Z \, D(c)Z \ldots D(c)^{k-1}Z] .$$

The matrices $V_k(c, Z)$ are the controllability matrices of the pair $(D(c), Z)$.

We recall from systems theory that the pair $(D(c), Z)$ is said to be controllable if the matrices $V_k(c, Z)$ have full rank for sufficiently large k.

Remark 2.1. $(D(c), Z)$ is controllable if and only if the vectors z_i corresponding to one and the same value c_i are linearly independent.

This follows from the fact that the subspace $\bigcap_{k=1}^{\infty} \ker V_k(c, Z)^T$ is invariant under the diagonal matrix $D(c)$.

We present an inversion formula for generalized Cauchy matrices. This formula is based on the fact that matrices C (such as many other types of structured matrices) are the solution of certain Sylvester matrix equations with a low rank right–hand side.

In fact, we have for a matrix 1.1

$$(2.1) \qquad D(c)C - CD(d) = ZY^T.$$

THEOREM 2.2. *Suppose that one of the following conditions is fulfilled:*
1) The pair $(D(c), Z)$ *is controllable and the equation*

$$(2.2) \qquad CX = Z$$

is solvable.
2) The pair is $(D(d), Y)$ *is controllable and the equation*

$$(2.3) \qquad W^T C = Y^T$$

is solvable.
Then C *is nonsingular and the inverse given by*

$$(2.4) \qquad C^{-1} = - \left[\frac{x_i^T w_j}{d_i - c_j} \right]_{i,j=1}^n .$$

where

$$X = \mathrm{col}\,(x_i^T)_1^n, \quad W = \mathrm{col}\,(w_j^T)_1^n.$$

Vice versa, if C *is nonsingular then 2.2 and 2.3 are solvable and* $(D(c), Z)$ *and* $(D(d), Y)$ *are controllable.*

Proof. Assume that the first condition is fulfilled. Then 2.2 implies

$$C(D(d) + XY^T) = D(c)C.$$

Hence

$$C(D(d) + XY^T)^k X = D(c)^k Z .$$

for $k = 0, 1, \dots$. The latter means that all columns of the matrices $V_k(c, z)$ belong to the range of C . Since $V_k(c, Z)$ has, by assumption, full range for sufficiently large k and C is square, the matrix C is nonsingular.

The proof is analogous if condition 2.3 is fulfilled.

Conversely, if C is nonsingular then clearly 2.2 and 2.3 are solvable. We show the controllability of $(D(c), Z)$. Assume, for a moment, that $(D(c), Z)$ is not controllable. Then according to Remark 2.1 there exists a set of indices J that the c_i for $i \in J$ coincide and the vectors

z_i $(i \in J)$ are linearly dependent. Suppose that the nontrivial combination $\sum_{i \in J} \alpha_i z_i$ vanishes. Setting $\alpha_i = 0$ for $i \in J$ and $\alpha = (\alpha_i)_1^n$ we have $\alpha^T C = 0$ which is a contradiction. $\qquad\square$

Let us discuss some computational viewpoints of formula 2.4. If the solutions X and W are known then 2.4 can be used to compute the solution of an equation $C\xi = \eta$ with $O(n^2)$ flops or $O(n)$ steps in parallel computation with n processors. But there is also the possibility to solve the equation with $O(n \log^2 n)$ computational amount if FFT is used. In fact it is well–known (see [12], [11], [20], [21], [22]) that the multiplication of a Cauchy matrix by a vector can be carried out with this complexity. A generalized Cauchy matrix 1.1 can be represented in the form

$$C = \sum_{k=1}^{r} D(z^k) C_0 D(y^k) \,,$$

where

$$z^k = (z_i^k)_{i=1}^n \quad, \qquad z_i = \operatorname{col}\,(z_i^k)_{k=1}^r$$
$$y^k = (y_i^k)_{i=1}^n \quad, \qquad y_i = \operatorname{col}\,(y_i^k)_{k=1}^r$$

and

$$C_0 = \left[\frac{1}{c_i - d_j} \right]_1^n \,.$$

Representing C^{-1} in analogous form the solution of $C\xi = \eta$ can be reduced to the multiplication of a Cauchy matrix by r vectors and $O(n)$ flops.

However, let us remark that the multiplication of Cauchy matrices by vectors with the help of FFT suffers from instable behaviour for large n in the general case (see [21] for an alternative computation). An exception is the case of unit roots c_i and d_j .

3. Interpolation interpretion and fundamental matrices. We give now an interpolation interpretation of the equation

$$(3.1) \qquad\qquad C\xi = \eta \qquad (\xi = (\xi_j)_1^n, \ \eta = (\eta_i)_1^n) \,.$$

For this we introduce into consideration vector functions of the form

$$(3.2) \qquad\qquad f(\lambda) = \sum_{j=1}^{n} \frac{\xi_j}{\lambda - d_j} y_j \,.$$

The function $f(\lambda)$ is proper rational and has a prescribed pole characteristics.

The following observation is obvious.

PROPOSITION 3.1. *The vector ξ is a solution of the equation 3.1 if and only if the function $f(\lambda)$ meets the interpolation conditions*

$$(3.3) \qquad z_i^T f(c_i) = \eta_i \quad (i = 1, \cdots, n) \ .$$

The interpolation problem in Prop. 3.1 is a simple case of a tangential (or directional) rational interpolation problem. For more information about this subject we refer to the recent monograph [3], which gives a fairly complete picture on the subject.

Now we give an interpolation interpretation of the equations 2.2 and 2.3.

PROPOSITION 3.2.

1) If the $r \times r$ matrix function

$$(3.4) \qquad \Phi(\lambda) = I_r - \sum_{j=1}^{n} \frac{1}{\lambda - d_j} y_j x_j^T$$

meets the interpolation conditions

$$(3.5) \qquad z_i^T \Phi(c_i) = 0 \quad (i = 1 \ldots, n)$$

then $X = \mathrm{col}\,(x_j^T)_1^n$ is a solution of 2.2. Vice versa, if X solves 2.2 then $\Phi(\lambda)$ satisfies 3.5.

2) If the $r \times r$ matrix function

$$(3.6) \qquad \Psi(\lambda) = I_r + \sum_{i=1}^{n} \frac{1}{\lambda - c_i} w_i z_i^T$$

meets the interpolation conditions

$$(3.7) \qquad \Psi(d_j) y_j = 0 \quad (j = 1, \cdots, n)$$

then $W = \mathrm{col}\,(w_i^T)_1^n$ is a solution of 2.3. Vice versa, if W solves 2.3 then $\Psi(\lambda)$ satisfies 3.7.

Proof. The conditions 3.5 can be written in the form

$$z_i^T = \sum_{j=1}^{n} \frac{1}{c_i - d_j} \, z_i^T y_j x_j^T \ .$$

This is equivalent to 2.2. Likewise, 3.7 is equivalent to 2.3.   □

The matrix function Φ will be called *right* and the matrix function $\Psi(\lambda)$ *left fundamental matrix* corresponding to the data (c, d, Z, Y) or to the matrix C. The concept of fundamental matrix will be generalized in our paper [16]. The fundamental matrices can be represented in the form

$$(3.8) \qquad \Phi(\lambda) = I_r - Y^T (\lambda I_n - D(d))^{-1} X$$

and

$$(3.9) \qquad \Psi(\lambda) = I_r + W^T(\lambda I_n - D(c))^{-1} Z \ .$$

In the language of linear systems theory, this means that the quatruple $[D(d),\ X,\ -Y^T,\ I_r]$ is a realization of $\Phi(\lambda)$ and $[D(c),\ Z,\ W^T,\ I_r]$ is a realization of $\Psi(\lambda)$ (see [2], [3]). In view of the controllability assumptions the realizations are minimal.

It is a well-known fact in realization theory that for $\Phi(\lambda)$ given by 3.8 one has

$$\Phi(\lambda)^{-1} = I_r + Y^T(\lambda I_n - D(d)^\times)^{-1} X \ ,$$

where $D(d)^\times$ is the so-called associated operator defined by

$$D(d)^\times = D(d) + XY^T \ .$$

From 2.1 we obtain

$$D(d) = C^{-1}D(c)C - XY^T \ .$$

Hence

$$D(d)^\times = C^{-1}D(c)C \ ,$$

which implies

$$\Phi(\lambda)^{-1} = I_r + Y^T C^{-1}(\lambda I_n - D(c))^{-1} CX \ .$$

Comparing this with 3.9 we obtain the following remarkable fact.

PROPOSITION 3.3. *If* $\Phi(\lambda)$ *is the right and* $\Psi(\lambda)$ *is the left fundamental matrix of* C *then*

$$\Psi(\lambda) = \Phi(\lambda)^{-1} \ .$$

for all $\lambda \neq d_j$.

4. Type I—algorithm. [1]

In this section we present a recursive procedure to compute the fundamental matrices of a Cauchy matrix C . This leads to an algorithm for the computation of the matrices X and W involved in the inversion formula 2.4 for C .

Together with the matrix C we consider the nested submatrices

$$C_k = \left[\frac{z_i^T y_j}{c_i - d_j} \right]_{i,j=1}^k \qquad (k = 1, \ldots, n) \ .$$

---

[1] The distinction of type I and type II algorithms was suggested e.g. in [5].

Then for each k for which C_k is nonsingular a right fundamental matrix Φ_k and a left fundamental matrix Ψ_k exist.

The recursion $k \to k+1$ is descibed in the following theorem.

THEOREM 4.1. *Suppose that Φ_k is a right and Ψ_k is a left fundamental matrix for C_k,*

$$(4.1) \qquad g_k := \Psi_k(d_{k+1})y_{k+1}, \quad h_k^T := z_{k+1}^T \Phi_k(c_{k+1}) .$$

Then C_{k+1} is nonsingular if and only if $h_k^T g_k \neq 0$, and fundamental matrices for C_{k+1} are given by

$$(4.2) \qquad \Phi_{k+1}(\lambda) = \Phi_k(x) \left(I_r - \frac{\alpha_k}{\lambda - d_{k+1}} g_k h_k^T \right)$$

$$(4.3) \qquad \Psi_{k+1}(\lambda) = \left(I_r + \frac{\alpha_k}{\lambda - c_{k+1}} g_k h_k^T \right) \Psi_k(\lambda) ,$$

where

$$(4.4) \qquad \alpha_k = \frac{c_{k+1} - d_{k+1}}{h_k^T g_k} .$$

Proof. Suppose that C_{k+1} is nonsingular. Then there exists a fundamental matrix for C_{k+1} which has the form

$$\Phi_{k+1}(\lambda) = I_r - Y_{k+1}^T (\lambda I_{k+1} - \widetilde{D})^{-1} \widetilde{X} ,$$

where $Y_{k+1} = \mathrm{col}\,(y_j)_1^{k+1}$, $\widetilde{D} = \mathrm{diag}\,(d_j)_1^{k+1}$ and $\widetilde{X}$ is a $(k+1) \times r$ matrix. We employ now realization theory (see [2]). The associated operator of $\widetilde{D}$ is given by $\widetilde{D}^\times = \widetilde{D} + Y_{k+1}^T \widetilde{X}$, and in view of 2.1 we have

$$\widetilde{D}^\times = C_{k+1}^{-1} \, \mathrm{diag}\,(c_i)_1^{k+1} C_{k+1} .$$

Since C_k is nonsingular the last component of the last column b of C_{k+1}^{-1} is nonzero. Hence the subspace spanned by b is angular with respect to the decomposition $\mathbf{C}^{k+1} = \mathcal{F}_1 \oplus \mathcal{F}_2$, where $\mathcal{F}_1$ denotes the subspace with vanishing last component and $\mathcal{F}_2$ the subspace with vanishing first k components. Thus Theorem 5.6 of [2] can be applied.

According to this theorem there exists a factorization

$$(4.5) \qquad \Phi_{n+1}(\lambda) = \Phi_n(\lambda) \left(I_r - \frac{\alpha_k}{\lambda - d_{k+1}} g_k h_k^T \right)$$

where c_{k+1} is a pole of $\left(I_r - \frac{\alpha_k}{\lambda - d_{k+1}} g_k h_k^T \right)^{-1}$. Since

$$\left(I_r - \frac{\alpha_k}{\lambda - d_{k+1}} g_k h_k^T \right)^{-1} = I_r + \frac{\alpha_k}{\lambda - d_{k+1} - \alpha_k h^T g_k} g_k h_k^T ,$$

we get $c_{k+1} = d_{k+1} = \alpha_k h_k^T g_k$ and

$$(4.6) \qquad \Psi_{n+1}(\alpha_k) = \left(I_r + \frac{\alpha_k}{\lambda - c_{k+1}} g_k h_k^T \right) \Psi_n(\alpha_k) \, .$$

In particular, $h^T g \neq 0$.

Inserting the interpolation conditions

$$z_{k+1}^T \, \Phi_{k+1}(c_{k+1}) = 0, \quad \Phi_{k+1}(d_{k+1}) y_{k+1} = 0$$

into 4.5 and 4.6 we obtain 4.1.

Assume now that $h_k^T g_k \neq 0$ and Φ_{k+1} and Ψ_{k+1} are defined by 4.2 and 4.3. Then it is easily checked that Φ_{k+1} and Ψ_{k+1} meet the interpolation conditions 3.5 and 3.7, respectively, where n has to be replaced by $k+1$. That means, the fundamental equations 2.2 and 2.3 are solvable for C_{k+1} . Hence C_{k+1} is nonsingular by Theorem 2.1.

Now we translate the recursions of Theorem 4.1 into matrix language. We introduce the matrices

$$Z_k := \operatorname{col} (z_i^T)_1^k, \quad Y_k := \operatorname{col} (y_j^T)_1^k \, .$$

Furthermore, let $X_k = \operatorname{col} (x_{ki}^T)_{i=1}^k$, $W_k = \operatorname{col} (w_{kj}^T)_{j=1}^k$ be the solutions of the equations

$$(4.7) \qquad C_k X_k = Z_k, \quad W_k^T C_k = Y_k^T$$

THEOREM 4.2. *The solutions of 4.7 satisfy the following recursions*

$$(4.8) \qquad X_{k+1} = \begin{bmatrix} X_k \\ 0 \end{bmatrix} - \alpha_k \begin{bmatrix} (D_k(d) - d_{k+1} I_k)^+ X_k g_k \\ 1 \end{bmatrix} h_k^T$$

$$(4.9) \qquad W_{k+1} = \begin{bmatrix} W_k \\ 0 \end{bmatrix} + \alpha_k \begin{bmatrix} (D_k(c) - c_{k+1} I_k)^+ W_k h_k \\ 1 \end{bmatrix} g_k^T \, ,$$

where

$$h_k^T = z_{k+1}^T - p^T X_k, \quad g_k = y_{k+1} - W_k^T q \, ,$$

p^T *denotes the last row and* q *the last column of* C_{k+1} *cancelling the last component,* α_k *is defined by 4.4,* $D_k(d) := \operatorname{diag} (d_i)_1^k$, *and the superscript "+" denotes the Moore-Penrose inverse*[2].

---

[2] We have $(D_k(d) - d_{k+1} I_k)^+ = \operatorname{diag} (d_i - d_{k+1}^+)_{i=1}^k$, where, for a number t, $t^+ :=$ $1/t$, if $t \neq 0$ and $t^+ := 0$ if $t = 0$.

Proof. According to 4.2 we have

$$(4.10) \quad \Phi_{k+1}(\lambda) = \left(I_r - \sum_{j=1}^{k} \frac{1}{\lambda - d_j} y_j x_{kj}^T \right) \left(I_r - \frac{\alpha_k}{\lambda - d_{k+1}} g_k h_k^T \right)$$

$$(4.11) \qquad = I_r - \sum_{d_j = d_{k+1}} \frac{1}{\lambda - d_j} y_j x_{kj}^T ,$$

where

$$(4.12) \qquad \Theta = I_r - \sum_{d_j \neq d_{k+1}} \frac{1}{d_{k+1} - d_j} y_j x_{kj}^T .$$

ince

$$\Theta g_k = \lim_{\lambda \to d_{k+1}} \Phi_k(\lambda) g_k ,$$

we get from 4.1 $\Theta g_k = y_{k+1}$. This implies the following relations

$$(4.13) \quad x_{k+1,j} = \begin{cases} x_{kj} - \alpha_k \frac{x_{kj}^T g_k}{d_j - d_{k+1}} h_k & : \quad d_j \neq d_{k+1}, \quad j = 1, \ldots, k \\ x_{kj} & : \quad d_j = d_{k+1}, \quad j = 1, \ldots, k \end{cases}$$

$$x_{k+1,k+1} = -\alpha_k h_k$$

$$w_{k+1,i} = \begin{cases} w_{ki} + \alpha_k \frac{h_k^T w_{ki}}{c_i - c_{k+1}} g_k & : \quad c_i \neq c_{k+1}, \quad i = 1, \ldots, k \\ w_{ki} & : \quad c_i = c_{k+1}, \quad i = 1, \ldots, k \end{cases}$$

$$w_{k+1,k+1} = -\alpha_k g_k .$$

The recursions 4.8 and 4.9 are just the matrix form of these relations.
$\square$

Let us estimate the number of multiplications of each step. We need twice kr multiplications in order to get h_k and g_k . To get $X_k g_k$ and $W_k h_k$ also twice kr multiplications are required. For the multiplication by the diagonal matrices we need $4k$ and by h_k^T and g_k^T another $2kr$ multiplications.

All together this gives $(6r + 4)k$. Summing up, we conclude that for the computation of the solutions of 2.2 and 2.3 about $(3r + 2)n(n - 1)$ multiplication are needed.

The algorithm described by the recursions 4.8 and 4.9 work if and only if $h_k^T g_k \neq 0$ for all $k = 1, \ldots, n$. This is equivalent to the strong nonsingularity of the matrix C . If, for a certain k, $h_k^T g_k = 0$ the algorithm breaks down. However one can proceed if one permutes some columns and/or rows of C. That means the pair (c_{k+1}, z_{k+1}) has to be replaced by a pair (c_l, z_l) and the pair (d_{k+1}, y_{k+1}) by a pair (d_m, y_m), where $l, m \in \{k+1, \ldots, n\}$. The integers l and m have to be chosen in such a way that

$$(4.14) \qquad (z_l^T - p^T X_k)(y_m - W_k^T q) \neq 0 .$$

Due to the nonsingularity of C, for each m there exists an l such that 4.14 is fulfilled.

In order to avoid instable behaviour of the algorithm it is recommended to make such an replacement also if $h_k^T g_k \neq 0$ but the number α_k is large. However, the search for the optimal m and l, i.e. those for which the corresponding $|\alpha_k|$ is minimal, will slow down the algorithm. Therefore, it would be desirable to have some criteria to decide in practical computation which α_k "good enough".

5. Type-II algorithm. The disadvantage of the algorithm described in Section 4 is that it is not very convenient for parallel computation since at each step the vectors g_k and h_k have to be calculated, which is in principle an inner product calculation and should be avoided. For parallel processing it is convenient to precompute these parameters. We show how this can be done. We introduce the vectors

$$(5.1) \qquad z_{ki}^T := z_i^T \Phi_k(c_i), \quad y_{kj} := \Psi_k(d_j) y_i$$

and the $n \times r$ matrices

$$(5.2) \qquad Z^k := \mathrm{col}\,(z_{ki}^T)_{i=1}^n, \ Y^k := \mathrm{col}\,(y_{kj}^T)_{j=1}^n .$$

Note that $z_{ki} = y_{ki} = 0$ for $i \leq k$. Furthermore, we observe that the vectors g_k and h_k appearing in the recursions of Theorem 4.1 and 4.2 are given by

$$(5.3) \qquad h_k = z_{k,k+1} \text{ and } g_k = y_{k,k+1} .$$

Now we obtain from Theorem 4.1 the following result.

THEOREM 5.1. *The matrices* Z^k *and* Y^k *fulfill the recursion*

$$(5.4) \qquad Z^{k+1} = Z^k - \alpha_k (D(c) - d_{k+1} I_n)^{-1} Z^k R_k$$

$$(5.5) \qquad Y^{k+1} = Y^k + \alpha_k (D(d) - c_{k+1} I_n)^{-1} Y^k R_k ,$$

where

$$R_k = y_{k,k+1} z_{k,k+1}^T$$

and

$$\alpha_k = \frac{c_{k+1} - d_{k+1}}{z_{k,k+1}^T y_{k,k+1}} .$$

Proof. Relation 4.2 implies

$$(5.6) \qquad z_{k+1,i}^T = z_{ki}^T - \frac{\alpha_k}{c_i - d_{k+1}} z_{ki}^T R_k \quad (i = 1, \ldots, n) .$$

The recursion 5.4 is just the matrix representation of 5.6. Analogously 5.5 is shown. $\square$

If we take now the recursions 4.8, 4.9 and 5.3, 5.4 together and take 5.2 into account we will get an algorithm which can completely be parallelized and carried out in $O(n)$ steps if n processors are available.

We continue with some remarks about pivoting. Of course, it is reasonable to reorder the data in such a way that $\left| \frac{z_1^T y_1}{c_1 - d_1} \right|$ is maximal or at least not too small.

We introduce the generalized Cauchy matrices

$$\widetilde{C}_k = [a_{ij}^k]_{i,j=k+1}^n := \left[\frac{z_{ki}^T y_{kj}}{c_i - d_j} \right]_{i,j=k+1}^n .$$

In particular, we have

$$(D(d) - c_{k+1} I_n)^{-1} \alpha_k = (a_{k+1,k+1}^k)^{-1} .$$

If k steps of the algorithm are done, i.e. if X_k and W_k are known, then the remaining steps are equivalent to the application of the algorithm to the matrix $\widetilde{C}_k$.

In order to guarantee stability one has to reorder $\widetilde{C}_k$ such that

$$\left| \frac{z_{k,k+1}^T y_{k,k+1}}{c_{k+1} - d_{k+1}} \right|$$

is maximal or at least not too small. A complete pivoting would require the computation of all entries of $\widetilde{C}_k$ which would slow down the algorithm essentially. Therefore, it is recommended to work with a partial pivoting.

Let us still note a recursion formula for the entries of $\widetilde{C}_k$.

PROPOSITION 5.2. *For the entries of the matrices* $\widetilde{C}_k$ *the following recursion holds*

$$(5.7) \qquad a_{ij}^{k+1} = a_{ij}^k - \frac{a_{i,k+1}^k a_{k+1,j}^k}{a_{k+1,k+1}^k} .$$

Formula 5.7 follows from 5.4 and 5.5 after an elementary calculation.

Relation 5.7 means nothig else that $\widetilde{C}_{k+1}$ is obtained from $\widetilde{C}_k$ by Gaussian elimination. It is well–known that Gaussian elimination with complete pivoting is stable and with partial pivoting is oostable in practice" (see e.g. [15]). Therefore, we may expect that the algorithm provided by Theorem 5.1 has the same stability properties if pivoting is used.

6. Divide-and-conquer approach. The complexity of algorithms for the solution of structured systems and for the solution of interpolation problems can be often reduced if one applies a divide–and–conquer strategy and FFT. For Toeplitz and Hankel matrices and the corresponding Padé approximation problems this was shown for example in [4], [9], [17], for Vandermonde and Cauchy matrices see [1] and [11]. All references above offer algorithms with computional amount $O(n \log^2 n)$ flops. Below we present an algorithm with complexity $O(n \log^3 n)$ for the computation of the matrices X and W, which leads together with the formula for C^{-1} to an algorithm with the same complexity for the solution of systems $C\xi = \eta$. The proposed algorithm seems to be of practical importance only in case the numbers c_i and d_j are unit roots. In this case the complexity can be reduced to $O(n \log^2 n)$. Actually generalized Cauchy matrices with c_i and d_j being unit roots just appear after transforming close–to–Toeplitz matrices into generalized Cauchy matrices (see Section 7 below). We also show in Section 8 that arbitrary generalized Cauchy matrices can be transformed into those with c_i and d_j being unit roots but with an increase of r by two.

The basic idea of the divide-and-conquer approach is to split the original problem into two subproblems of the same structure and about half the size. In our situation we compute first the fundamental matrices Φ_m and Ψ_m for $m \approx \frac{n}{2}$.

Then we determine the matrices Z^m and W^m with the columns $z_{mi}^T = z_c^T \Phi_m(c_i)$ and $w_{mi}^T = w_i^T \Psi_m(d_i)$ (the first m columns of these matrices vanish!). This is equivalent to the multiplication of X_m and W_m by the generalized Cauchy matrices

$$C_m^1 = \left[\frac{z_i^T y_j}{c_i - d_j} \right]_{i=m+1 \; j=1}^{n \qquad m} \quad \text{and} \quad C_m^2 = \left[\frac{z_j^T y_i}{c_j - d_i} \right]_{i=m+1 \; j=1}^{n \qquad m},$$

respectively, since

$$Z - C \begin{bmatrix} X_m \\ 0 \end{bmatrix} = Z^m \quad \text{and} \quad W - C^T \begin{bmatrix} W_m \\ 0 \end{bmatrix} = Y^m.$$

As remarked in Section 2, this multiplication can be carried out with $O(n \log^2 n)$ flops.

As the next step, we seek rational matrix functions $\widetilde{\Phi}_m$ and $\widetilde{\Psi}_m$ with poles at d_i, c_i ($i = m+1, \ldots, n$), respectively, meeting the interpolation conditions

$$z_{mi}^T \widetilde{\Phi}_m(c_i) = 0 \quad \text{and} \quad \widetilde{\Psi}_m(d_i) y_{mi} = 0 \quad (i = m+1, \ldots, n) \,.$$

In other words, we determine the solutions of the equations

$$\widetilde{C}_m \widetilde{X}_m = \widetilde{Z}^m \quad \text{and} \quad \widetilde{W}_m^T \widetilde{C}_m = (\widetilde{Y}^m)^T \,,$$

where

$$\widetilde{C}_m = \left[\frac{z_{mi}^T y_{mi}}{c_i - d_j} \right]_{i,j=m+1}^n$$

$$\widetilde{Z}^m = \operatorname{col} (z_{mi}^T)_{i=m+1}^n, \quad \widetilde{Y}^m = \operatorname{col} (y_{mi}^T)_{i=m+1}^n \,.$$

Now we have

$$\Phi = \Phi_m \widetilde{\Phi}_m \quad \text{and} \quad \Psi = \widetilde{\Psi}_m \Psi_m \,.$$

These multiplication could be carried out with FFT and an amount of $O(n \log n)$, provided that the entries of Φ and Ψ are given as rational functions. However, this multiplication is not necessary in order to get the matrices X and W.

In order to compute X and W let us assume, for simplicity that the numbers c_i and d_i are pairwise different. In this case the residue of Φ at the equals $-y x_i^T$ and the residue of Ψ at c_i equals $w_i z_i^T$. Hence we have

$$y_i x_i^T = \begin{cases} y_i x_{mi}^T \widetilde{\Phi}_m(d_i) & : \quad i = 1, \ldots, m \\ \Phi_m(d_i) y_{mi} \widetilde{X}_{mi}^T & : \quad i = m+1, \ldots, n \end{cases} \,,$$

$$w_i z_i^T = \begin{cases} \widetilde{\Psi}_m(c_i) w_{mi} z_i^T & : \quad i = 1, \ldots, m \\ \widetilde{W}_{mi} z_{mi}^T \Psi_m(c_i) & : \quad i = m+1, \ldots, n \end{cases} \,,$$

where the $\widetilde{X}_{mi}^T$ and $\widetilde{W}_{mi}^T$ are the rows of $\widetilde{X}_m$ and $\widetilde{W}_m$, respectively. Consequently,

$$x_i^T = \begin{cases} x_{mi}^T \widetilde{\Phi}_m(d_i) & : \quad i = 1, \ldots, m \\ \widetilde{X}_{mi}^T & : \quad i = m+1, \ldots, n \end{cases} \,,$$

$$w_i = \begin{cases} \widetilde{\Psi}_m(c_i) w_{mi} & : \quad i = 1, \ldots, m \\ \widetilde{W}_{mi} & : \quad i = m+1, \ldots, n \end{cases} \,.$$

That means in order to compute x_i and w_i it remains to determine $\widetilde{\Phi}_m(d_i)$ and $\widetilde{\Psi}_m(c_i)$ $(i = 1, \ldots, m)$ which is again a multiplication of vectors by a generalized Cauchy matrix.

Let $A(n)$ denote the computational amount to determine Φ_n and Ψ_n. Then the procedure described above admits the estimation

$$A(n) = 2A\left(\frac{n}{2}\right) + O(n\log^2 n),$$

which implies

$$A(n) = O(n\log^3 n).$$

Consider now the special case of unit roots

$$c_i = \exp\frac{2i\pi}{n}\sqrt{-1}, \quad d_j = \exp\frac{(2j+1)\pi}{n}\sqrt{-1},$$

which just occur in Section 7. Note that $c_i^n = 1$ and $d_j^n = -1$. In this case the fundamental matrix Φ can be represented in the form

$$\Phi(x) = \frac{1}{\lambda^n + 1}P(\lambda),$$

where $P(\lambda)$ is monic $r \times r$ polynomial matrix with degree n.

That means the steps in the algorithm described above reduce to multiplication of matrix polynomials, evaluation of values at unit roots and interpolation at unit roots. Applying FFT the amount will be reduced to $O(n\log^2 n)$.

7. Application to other types of structured matrices. Let U and V be two fixed $n \times n$ matrices. A matrix A is said to possess a *(Sylvester) UV–displacement structure* if $r := \text{rank}(VA - AU)$ is small compared with n. The integer r is called the (Sylvester) UV–displacement rank. This concept was introduced in [18] generalizing the displacement concept of T. Kailath et al. ([19]; see also [7] and references therein). By definition, generalized Cauchy matrices are just the matrices with an $(D(c), D(d))$–displacement structure.

It is an elementary but possibly important fact that in many cases matrices with a UV–displacement structure can be transformed into generalized Cauchy matrices. Let us show this.

We assume that U and V are of simple structure. i.e. possess diagonalizations

$$(7.1) \qquad U = Q^{-1}D(d)Q, \quad V = R^{-1}D(c)R,$$

where $d = (d_j)_1^n$, $c = (c_i)_1^n$. Furthermore we assume that U and V have no eigenvalues in common, i.e. $c_i \neq d_j$ if $i \neq j$. Then the following is easily verified.

PROPOSITION 7.1. *Suppose that*

$$VA - AU = GH^T,$$

where $G, H \in \mathbf{C}^{n \times r}$, and $RG = \operatorname{col}(z_i^T)_1^n$, $Q^{-T}H = \operatorname{col}(y_j^T)_1^n$ Then $C := RAQ^{-1}$ is a generalized Cauchy matrix and given by

$$C = \left[\frac{z_i^T y_j}{c_i - d_j} \right]_{i,j=1}^n .$$

Of course, in order to get some advantage of the transformation Q and R must have a simple form. Fortunately, in important cases this is fulfilled.

Let S_+ denote the cyclic and S_- the anticyclic forward shifts,

$$S_+ = \begin{bmatrix} 0 & \cdots & 0 & 1 \\ 1 & & 0 & 0 \\ & \ddots & & \vdots \\ & & 1 & 0 \end{bmatrix}, \quad S_- = \begin{bmatrix} 0 & \cdots & 0 & -1 \\ 1 & & 0 & 0 \\ & \ddots & & \vdots \\ & & 1 & 0 \end{bmatrix}$$

These matrices can be diagonalized with the help of the Fourier matrices $F_\pm$ defined as follows. Denote $\theta = \exp\left(-\frac{2\pi}{n}\sqrt{-1}\right)$, $\sigma = \exp\frac{\pi}{n}\sqrt{-1}$, $\theta_k^+ = \theta^k$ and $\theta_k^- = \theta^k \sigma$. The $\theta_k^\pm$ are the n-th roots of ± 1. Define $F_+ := \frac{1}{\sqrt{n}}[(\theta_j^+)^k]_{j,k=0}^{n-1}$ and $F_- := \frac{1}{\sqrt{n}}[(\theta_j^-)^k]_{j,k=0}^{n-1}$.

Note that F_+ is symmetric and unitary. Hence $F_+^{-1} = \overline{F}_+$. Furthermore,

$$F_- = F_+ \operatorname{diag}(1, \sigma, \ldots, \sigma^{-1})$$

The diagonalizations of $S_\pm$ are given by

$$S_+ = F_+^{-1} D_+ F_+, \quad S_- = F_-^{-1} D_- F_- ,$$

where

$$D_+ = \operatorname{diag}(\theta_0^+, \ldots, \theta_{n-1}^+)^{-1} \quad \text{and} \quad D_- = \operatorname{diag}(\theta_0^-, \ldots, \theta_{n-1}^-)^{-1} .$$

Now we consider matrices with UV–displacement structure if $U = S_+$ and $V = S_-$. Standard examples of a matrices with such a structure are Toeplitz matrices. In fact, let $T = [a_{i-j}]_1^n$. Then

$$(7.2) \qquad\qquad S_- T - T S_+ = GH^T ,$$

where

$$G^T = \begin{bmatrix} a_0 & a_1 - a_{1-n} & \cdots & a_{n-1} - a_{-1} \\ 1 & 0 & \cdots & 0 \end{bmatrix}$$

and

$$H^T = \begin{bmatrix} 0 & \cdots & 0 & 1 \\ a_{-1} + a_{n-1} & \cdots & a_{1-n} + a_1 & a_0 \end{bmatrix} .$$

That means Toeplitz matrices have displacement rank less or equal two.

For this reason matrices with small (S_+, S_-)–displacement rank will be referred to as close-to-Toeplitz and the rank of $S_- A - A S_+$ will be called Toeplitz-displacement rank [3].

By Proposition 7.1 $F_- A \overline{F}_+$ is a generalized Cauchy matrix if A is close-to-Toeplitz. For the case of a Toeplitz matrix we obtain in particular the following.

PROPOSITION 7.2. Let $T = [a_{i-j}]_1^n$ be a Toeplitz matrix. Then $C = [c_{ij}]_1^n := F_- T \overline{F}_+$, is a generalized Cauchy matrix given by

$$c_{ij} = \frac{z_i \theta^j + y_j}{\theta_i^+ - \theta_j^+} \, ,$$

where

$$z = (z_i)_0^{n-1} = F_- g, \quad y = (y_j)_0^{n-1} = \overline{F}_+ h \, ,$$

$$g := (a_i - a_{i-n})_0^{n-1}, \quad h := (a_{-i-1} + a_{n-i-1})_0^{n-1} \quad (a_n = a_{-n} = 0) \, .$$

Note that the matrix $L = C \mathrm{diag}\,(1, \theta, \ldots, \theta^{n-1})^{-1}$ is a Loewner matrix with entries

$$l_{ij} = \frac{z_i - y_j \theta^{-j}}{\theta_i^+ - \theta_j^-} \, .$$

8. Transformation to unit roots.

Among all generalized Cauchy matrices those with c_i and d_j being unit roots seem to be the most convenient ones. This concerns both stability and complexity matters. We show that arbitrary generalized Cauchy matrices can be transformed into those with this property.

Suppose that the matrix C is given by 1.1 with $c_i, d_j \neq 0$. Define

$$p(\lambda) := \prod_{i=1}^n (\lambda - c_i), \quad q(\lambda) := \prod_{j=1}^n (\lambda - d_j)$$

and introduce the companion matrix

$$B(p) := \begin{bmatrix} -p_{n-1} & \cdots & & -p_0 \\ 1 & & & \\ & \ddots & & \\ & & 1 & \end{bmatrix} ,$$

---

[3] The concepts of Toeplitz displacement rank in the literature are slightly different.

where $p(\lambda) := p_0 + p_1\lambda + \ldots + \lambda^n$, and the analogously defined matrix $B(q)$. Let furthermore V_c denote the Vandermonde matrix $V_c := [c_i^{-j+1}]_{i,j=1}^n$ and V_d the analogous matrix for d . Then

$$(8.1) \qquad B(p)V_c = V_c D(c) \quad \text{and} \quad B(q)V_d = V_d D(d) .$$

From 2.1 we obtain now

$$(8.2) \qquad V_c^{-1}B(p)V_c C - CV_d^{-1}B(p)V_d = ZY^T .$$

Next we note that

$$(8.3) \qquad B(p) = S_- - e_1\widetilde{p}^T \quad \text{and} \quad B(q) = S_+ - e_1\widetilde{q}^T .$$

for vectors $\widetilde{p}$ and $\widetilde{q}$, where e_1 is the first unit vector. We conclude that the matrix

$$(8.4) \qquad A := V_c C V_d^{-1}$$

fulfills an equation

$$(8.5) \qquad S_- A - AS_+ = V_c ZY^T V_d^{-1} + T ,$$

where T is a matrix with rank $T \leq 2$.

From 8.5 and Proposition 7.1 we conclude the following.

PROPOSITION 8.1. *If C is a generalized Cauchy matrix of the form 1.1 then A defined by 8.4 is close-to-Toeplitz with Toeplitz displacement rank $r+2$ at most.*

Furthermore the matrix

$$\widetilde{C} := F_- V_c C V_d^{-1} F_+^{-1}$$

is a generalized Cauchy matrix of the form

$$\widetilde{C} = \left[\frac{\widetilde{z}_i^T \widetilde{y}_j}{\theta_i^+ - \theta_j^-} \right]_1^n ,$$

where $\widetilde{z}_i, \widetilde{y}_j \in \mathbf{C}^{r+2}$ and $\theta_i^\pm$ are the n-th roots of ± 1 .

According to the considerations in Section 6 the evaluation of the fundamental matrices requires in the case of unit roots $O(n \log^2 n)$ operations. Furthermore, it is well–known that the multiplication of a Vandermonde matrix or its inverse by a vector can be carried out with the same complexity. As a consequence we obtain the following.

THEOREM 8.2. *Systems of equation with an $n \times n$ generalized Cauchy coefficient matrix can be solved with $O(n \log^2 n)$ complexity.*

REFERENCES

[1] A.V.AHO, J.E.HOPCROFT, J.D.ULLMAN, *The design and analysis of computer algorithms*, Addison-Wesley 1976.

[2] H.BART, I.GOHBERG, M.A.KAASHOEK, *Minimal factorization of matrix and operator functions*, Birkhäuser Verlag, Basel-Boston-Stuttgart 1979.

[3] J.A.BALL, I.GOHBERG, L.RODMAN, *Interpolation of rational matrix functions*, Birkhäuser Verlag, Basel-Boston-Stuttgart 1990.

[4] R.P.BRENT, F.G.GUSTAVSON, D.Y.Y.YUN, *Fast solution of toeplitz systems of equations and computation of the pade approximation*, J. Algorithmus 1 (1980), 259–295.

[5] A.BULTHEEL, *Laurent series and their padé approximation*, Birkhäuser Verlag, Basel-Boston-Stuttgart 1987.

[6] T.CHAN, C.HANSEN, *A look-ahead levinson algorithm for general toeplitz systems*, (to appear).

[7] J.CHUN, T.KAILATH, *Displacement structure for hankel, vandermonde, and related (derived) matrices*, Linear Algebra Appl. 151 (1991), 199–227.

[8] G.CYBENKO, *The numerical stability of the levinson-durbin algorithm for toeplitz systems of equations*, SIAM J. Sci. Stat. Comp. 1 (1980), 303–319.

[9] F.DE HOOG, *A new algorithm for solving toeplitz systems of equations*, Linear Algebra Appl. 88/89 (1987), 123–138.

[10] W.F.DONOGHUE, *Monotone matrix functions and analytic continuation*, Springer-Verlag, Berlin-Heidelberg-New York 1974.

[11] T.FINCK, G.HEINIG, K.ROST, *An inversion formula and fast algorithms for cauchy-vandermonde matrices*, Linear Algebra Appl. (to appear).

[12] A.GERASOULIS, M.D.GRIGORIADIS, L.SUN, *A fast algorithm for trummer's problem* SIAM J. Stat. Comp. 8 (1) 135–138.

[13] I.GOHBERG, I.KOLTRACHT, P.LANCASTER, *Efficient solution of linear systems of equations with recursive structure* Linear Algebra Appl. 80 (1986), 81–113.

[14] I.GOHBERG, T.KAILATH, I.KOLTRACHT, P.LANCASTER, *Linear complexity parallel algorithms for linear systems of equations with recursive structure* Linear Algebra Appl. 88/99 (1987), 271–316.

[15] G.GOLUB, C.F.VAN LOAN, *Matrix computations*, John Hopkins 1989.

[16] G.HEINIG, *Inversion of generalized cauchy and cauchy-vandermonde matrices* (in preparation).

[17] G.HEINIG, P.JANKOWSKI, *Parallel and superfast algorithms for hankel systems of equations*, Numerische Math. 58 (1990), 109–127.

[18] G.HEINIG, K.ROST, *Algebraic methods for toeplitz-like matrices and operators*, Birkhäuser Verlag, Basel-Boston-Stuttgart 1984.

[19] T.KAILATH, S.Y.KUNG, M.MORF, *Displacement rank of matrices and linear equations*, Journal of Math. Anal. Appl. 68 (1979), 395–407.

[20] V.PAN, *On computations with dense structured matrices*, Math. Comp. 55 (1990), 179–190.

[21] V.PAN, *Complexity of computations with matrices and polynomials*, SIAM Review 34 (2) (1992), 225–262.

[22] Z.VAVŘIN, *Remarks on complexity of polynomial and special matrix computations*, Linear Algebra Appl. 122/123/124 (1989), 539–564.

WAVELETS, FILTER BANKS, AND ARBITRARY TILINGS OF THE TIME-FREQUENCY PLANE*

C. HERLEY[†], J. KOVAČEVIĆ[‡], AND M. VETTERLI[§]

Abstract. Recent work has made it clear that the design of multirate filter banks for signal processing, and of wavelet bases for the analysis of functions, address essentially two versions of the same problem: construction of structured bases for the linear expansion of signals. In the filter bank case the signals are elements of some sequence space, while in the wavelet case they are from some function space, but the objectives, and designs in both cases are very similar. This paper reviews some of the recent developements in these fields.

Key words. Wavelets; Filter banks

1. Linear expansions of discrete-time signals. The fundamental problem with which we are concerned is the choice of bases for the linear expansion of signals. That is, given a discrete-time signal $x(n)$ we wish to find $a_i(n)$ and $b_i(n)$ such that we can write

$$(1.1) \qquad x(n) = \sum_i < x(n), a_i(n) > b_i(n).$$

If $b_i(n) = a_i(n)$ then (1.1) is the familiar orthonormal basis expansion formula [16]. Otherwise the $b_i(n)$ are a set of biorthogonal functions with the property

$$< b_j(n), a_i(n) > = \delta_{i-j}.$$

The function δ is defined such that $\delta_{i-j} = 0$, unless $i = j$, in which case $\delta_0 = 1$. We shall consider cases where the summation in (1.1) is infinite, but restrict our attention to the case where it is finite for the moment; that is, where we have a finite number N of data samples, and so the space is finite dimensional.

Assume that we are operating in C^N, and that we have N basis vectors, the minimum number to span the space. Since the transform is linear it can be written as a matrix. That is, if the $\mathbf{a}_i^*$ are the rows of a matrix $\mathbf{A}$,

---

* Work supported in part by the National Science Foundation under grants ECD-88-11111 and MIP 90-14189.

† The author is with Hewlett-Packard Laboratories, Palo Alto, CA 94304.

‡ The author is with AT&T Bell Laboratories, Murray Hill, NJ 07974.

§ The author is with the EECS Dept., University of California, Berkeley, CA 94720.

then

$$(1.2) \qquad \mathbf{A} \cdot \mathbf{x} = \begin{bmatrix} < x(n), a_0(n) > \\ < x(n), a_1(n) > \\ \vdots \\ < x(n), a_{N-2}(n) > \\ < x(n), a_{N-1}(n) > \end{bmatrix},$$

and if $\mathbf{b}_i$ are the columns of $\mathbf{B}$ then

$$(1.3) \qquad \mathbf{x} = \mathbf{B} \cdot \mathbf{A} \cdot \mathbf{x}.$$

Obviously, $\mathbf{B} = \mathbf{A}^{-1}$; if $\mathbf{B} = \mathbf{A}^*$ then $\mathbf{A}$ is unitary, $b_i(n) = a_i(n)$ and we have that (1.1) is the orthonormal basis expansion.

Clearly the construction of bases is not difficult: any nonsingular $N \times N$ matrix will do for this space. Similarly, to get an orthonormal basis we need merely take the rows of any unitary $N \times N$ matrix, for example the identity $\mathbf{I}_N$. Obviously we have considerable freedom in the choice of a basis, and additional constraints will come from requirements imposed by the application.

In signal processing a major application is signal compression, where we wish to quantize the input signal in order to transmit it with as few bits as possible, while minimizing the distortion introduced. If the input vector comprises samples of a real signal, then the samples are probably highly correlated, and the identity basis (where the ith vector contains a 1 in the ith position and is zero elsewhere) with scalar quantization will end up using many of its bits to transmit information which does not vary much from sample to sample. If we can choose a matrix $\mathbf{A}$ such that the elements of $\mathbf{A} \cdot \mathbf{x}$ are much less correlated than those of $\mathbf{x}$ then the job of efficient quantization becomes a great deal simpler [15]. In fact the Karhunen-Loève transform, which produces uncorrelated coefficients, is known to be optimal for fine quantization in a mean squared error sense [15], [36].

Since in (1.1) the signal is written as a superposition of the basis sequences $b_i(n)$ we can say that if $b_i(n)$ has most of its energy concentrated around time $n = n_0$ then the coefficient $< x(n), a_i(n) >$ measures to some degree the concentration of $x(n)$ at time $n = n_0$. Equally, taking the discrete Fourier transform of (1.1)

$$X(k) = \sum_i < x(n), a_i(n) > B_i(k),$$

we can see that if $B_i(k)$ has most of its energy concentrated about frequency $k = k_0$ then $< x(n), a_i(n) >$ measures to some degree the concentration of $X(k)$ at $k = k_0$. This basis function is mostly localized about the point (n_0, k_0) in the discrete-time discrete-frequency plane. Similarly, for each of the basis functions $b_i(n)$ we can find the area of the discrete-time discrete-frequency plane where most of their energy lies. All of the basis functions together will effectively cover the plane, since if any part were not covered there would be a "hole" in the basis, and we would not be able to completely represent all sequences in the space. Similarly the localization areas, or tiles, corresponding to distinct basis functions should not overlap by too much, since this would represent redundancy in the system.

Choosing a basis can then be loosely thought of as choosing some tiling of the discrete-time discrete-frequency plane. For example, Figure 1.1 shows the tiling corresponding to various orthonormal bases in C^{64}. The horizontal axis represents discrete time, and the vertical axis discrete frequency. Naturally, each of the diagrams contains 64 tiles, since this is the number of vectors required for a basis, and each tile can be thought of as containing 64 points out of the total of 64^2 in this discrete-time discrete-frequency plane. The first is the identity basis, which has narrow vertical strips as tiles, since the basis sequences $\delta(n+k)$ are perfectly localized in time, but have energy spread equally at all discrete frequencies. That is, the tile is one discrete-time point wide and 64 discrete-frequency points long. The second, shown in Figure 1.1(b), corresponds to the discrete Fourier transform basis vectors $e^{j2\pi in/N}$; these of course are perfectly localized at the frequencies $i = 0, 1, \cdots N - 1$, but have equal energy at all times (*i.e* 64 points wide, one point long). Figure 1.1(c) shows the tiling corresponding to a discrete orthogonal wavelet transform (or octave-band subband coder) operating over a finite-length signal. Figure 1.1(d) shows the tiling corresponding to a time-varying discrete orthogonal wavelet-packet transform operating over a finite-length signal, with arbitrary splits in time and frequency; construction of such schemes is discussed in Section 4. In Figure 1.1(c) and (d) the tiles have varying shapes, but still contain 64 points each.

It should be emphasized that the localization of the energy of a basis function to the area covered by one of the tiles is only approximate. Nonetheless the idea of considering the construction of bases from the point of view of time-frequency tilings is a very valuable one, since finding a basis that matches the time-frequency energy content of the signal holds considerable promise for compression.

In practice, of course, we will always deal with real signals, and in general we will restrict the basis functions to be real also. When this is so $\mathbf{B}^* = \mathbf{B}^T$ and the basis is orthonormal provided $\mathbf{A}^T\mathbf{A} = \mathbf{I} = \mathbf{A}\mathbf{A}^T$. Of the bases shown in Figure 1.1 only the discrete Fourier transform will be excluded with this restriction. One can, however, consider a real transform which has many properties in common with the DFT, for example the discrete Hartley transform [3].

While the above description was given in terms of finite-dimensional signal

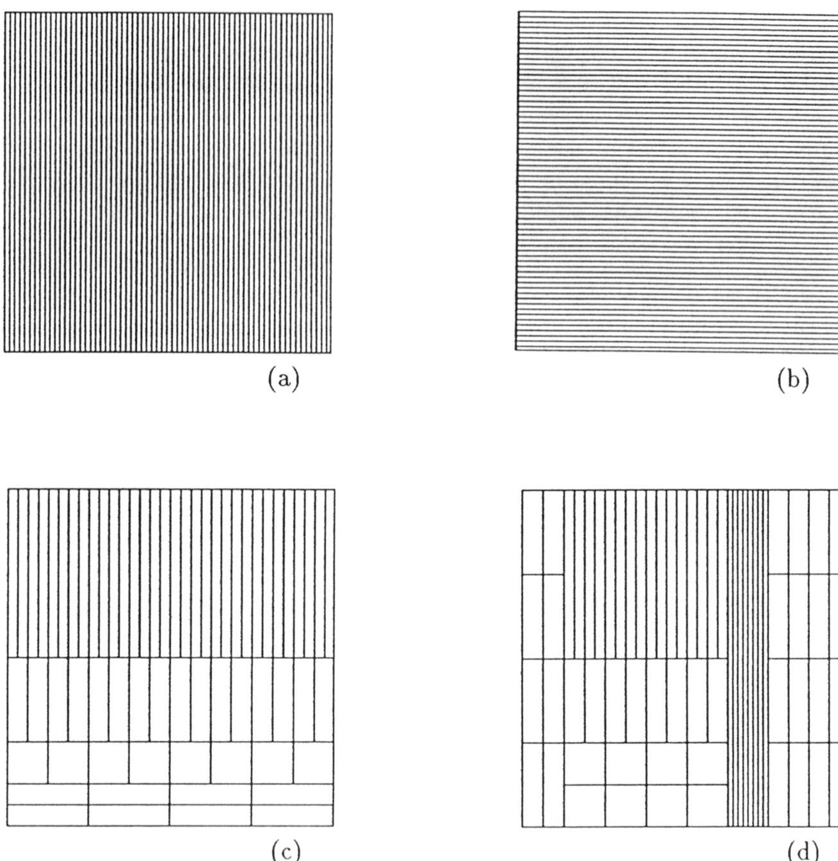

FIG. 1.1. *Examples of tilings of the discrete-time discrete-frequency plane; time is the horizontal axis, frequency the vertical. (a) The identity transform. (b) Discrete Fourier transform. (c) Finite-length discrete wavelet ransform. (d) Arbitrary finite length transform.*

spaces the interpretation of the linear transform as a matrix operation, and the tiling approach remain essentially unchanged in the case of infinite-length discrete-time signals. In fact for bases with the structure we desire, construction in the infinite-dimensional case is easier than in the finite-dimensional case. The modifications necessary for the transition from R^N to $l^2(Z)$ are that an infinite number of basis functions is required instead of N, the matrices $\mathbf{A}$ and $\mathbf{B}$ become doubly infinite, and the tilings are in the discrete-time continuous-frequency plane (the time axis ranges over Z, the frequency axis goes from 0 to π, assuming real signals).

Good decorrelation is one of the important factors in the construction of bases. If this were the only requirement we would always use the Karhunen-Loève transform, which is an orthogonal data-dependent transform which produces uncorrelated samples. This is not used in practice, since finding the coefficients of the matrix $\mathbf{A}$ can be difficult. Very significant also, however, is the complexity of calculating the coefficients of the transform using (1.2), and of putting the signal back together using (1.3). In general, for example, using the basis functions for R^N, evaluating each of the matrix multiplications in (1.2) and (1.3) will require $O(N^2)$ floating point operations, unless the matrices have some special structure. If, however, $\mathbf{A}$ is sparse, or can be factored into matrices which are sparse, then the complexity required can be dramatically reduced. This is the case with the discrete Fourier transform, where there is an efficient O(N logN) algorithm to do the computations, which has been responsible for its popularity in practice. This will also be the case with the transforms that we consider, $\mathbf{A}$ and $\mathbf{B}$ will always have special structure to allow efficient implementation.

This paper is intended to give a survey of recent results in the area, and is based largely on the material in the following references: [18], [19], [20], [21], [22], [24]. Section 2 examines the construction of two-channel filter banks in detail, and shows how these discrete-time bases can be used to generate continuous-time ones. Section 3 summarizes the state of multidimensional filter bank and wavelet design. In Section 4 we explore more general structures, where the analysis and synthesis structures are time-varying; this leads to the construction of bases with essentially arbitrary tiling patterns in the time-frequency plane.

2. Filter banks and wavelets. The methods of designing bases that we will employ draw on ideas first used in the construction of multirate filter banks. The idea of such systems is to take an input system an split it into subsequences using banks of filters. This simplest case involves splitting into just two parts using a structure such as shown in Figure 2.1. This technique has a long history of use in the area of subband coding: firstly of speech [9], [8] and more recently of images [52], [46]. In applied mathematics and computer vision, it appeared as wavelets and multiresolution analysis [11], [28]. Texts on the subject are [29], [43], [49].

We will consider only the two-channel case in this section. If $\hat{X}(z) = X(z)$ the filter bank has the perfect reconstruction property. Note that the z-transform is defined as

$$X(z) = \sum_{n=-\infty}^{\infty} x(n)z^{-n}.$$

It is easily shown that the output $\hat{X}(z)$ of the overall analysis/synthesis system is given by:

$$(2.1) \quad \hat{X}(z) \quad = \quad \frac{1}{2}[G_0(z)\, G_1(z)] \begin{bmatrix} H_0(z) & H_0(-z) \\ H_1(z) & H_1(-z) \end{bmatrix} \begin{bmatrix} X(z) \\ X(-z) \end{bmatrix}$$

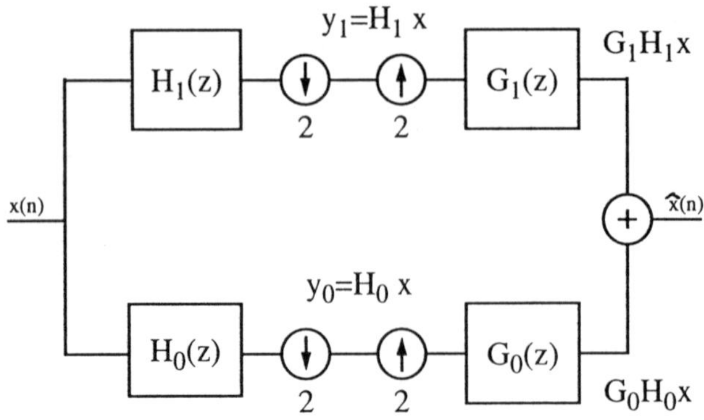

FIG. 2.1. *Maximally decimated two-channel multirate filter bank.*

$$= \frac{1}{2}[H_0(z)G_0(z) + H_1(z)G_1(z)] \cdot X(z)$$

$$+ \frac{1}{2}[H_0(-z)G_0(z) + H_1(-z)G_1(z)] \cdot X(-z).$$

We have used the fact that if $X(z)$ is processed by a subsampler followed by an upsampler the result contains the odd-indexed coefficients $1/2[X(z) + X(-z)]$. Call the 2×2 matrix $\mathbf{H}_m(z)$. This gives that the unique choice for the synthesis filters is

$$\begin{bmatrix} G_0(z) \\ G_1(z) \end{bmatrix} = \begin{bmatrix} H_0(z) & H_0(-z) \\ H_1(z) & H_1(-z) \end{bmatrix}^{-1} \cdot \begin{bmatrix} 2 \\ 0 \end{bmatrix}$$

$$(2.2) \qquad = \frac{2}{\Delta_m(z)} \begin{bmatrix} H_1(-z) \\ -H_0(-z) \end{bmatrix},$$

where $\Delta_m(z) = \det \mathbf{H}_m(z)$.

If we observe that $\Delta_m(z) = -\Delta_m(-z)$ and define

$$P(z) = \frac{2 \cdot H_0(z)H_1(-z)}{\Delta_m(z)} = H_0(z)G_0(z),$$

it follows from (2.2) that

$$G_1(z)H_1(z) = \frac{2 \cdot H_1(z)H_0(-z)}{\Delta_m(-z)} = P(-z).$$

We can then write that the necessary and sufficient condition for perfect reconstruction (2.1) is:

$$(2.3) \qquad P(z) + P(-z) = 2.$$

(a) $H_0(z)G_0(z)$

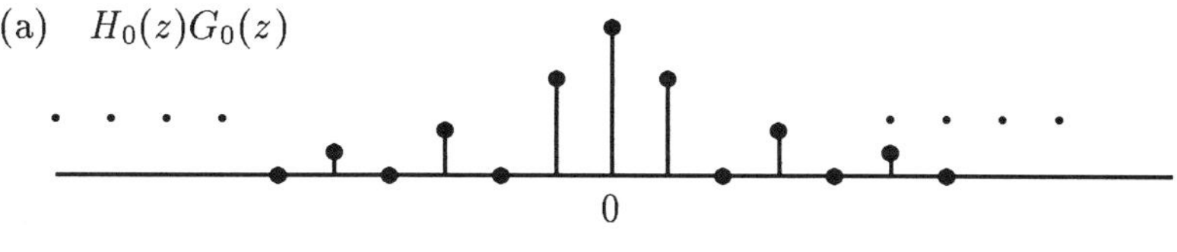

(b) $H_0(z)G_1(z)$

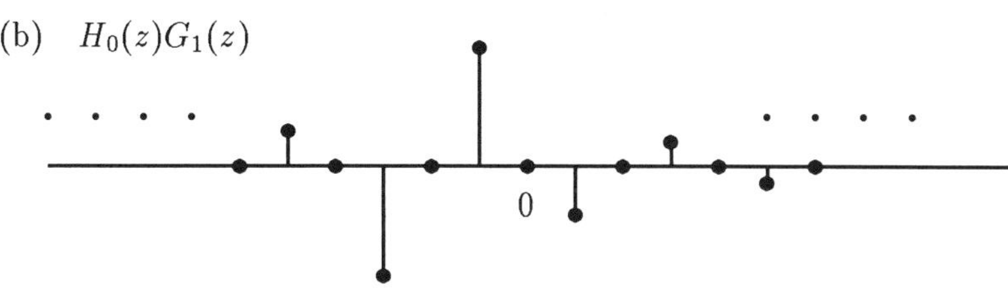

FIG. 2.2. *Zeros of the correlation functions. (a) Correlation $H_0(z)G_0(z)$. (b) Correlation $H_0(z)G_1(z)$.*

Since this condition plays an important role in what follows, we will refer to any function having this property as *valid*. The implication of this property is that all but one of the even-indexed coefficients of $P(z)$ are zero. That is

$$P(z) + P(-z) = \sum_n (p(n)z^{-n} + p(n)(-z)^{-n})$$
$$= \sum_n 2 \cdot p(2n)z^{-(2n+1)}.$$

For this to satisfy (2.3) requires $p(2n) = \delta_n$; thus only one of the even-indexed samples of $P(z)$ is non-zero. Such a function is illustrated in Figure 2.2(a).

Constructing such a function is not difficult. In general, however, we will wish to impose additional constraints on the filter banks. So $P(z)$ will have to satisfy other constraints, in addition to (2.3).

Observe that, as a consequence of (2.2) $G_0(z)H_1(z)$, *i.e.* the cross-correlation of $g_1(n)$ and the time-reversed filter $h_1(-n)$, and $G_1(z)H_0(z)$, the cross-correlation of $g_1(n)$ and $h_0(-n)$, have only odd-indexed coefficients, just as for the function in Figure 2.2(b), that is:

(2.4) $$< g_0(n), h_1(2k - n) > = 0,$$

$$(2.5) \qquad\qquad < g_1(n), h_0(2k - n) > \; = \; 0,$$

(note the time reversal in the inner product). Define now the matrix $\mathbf{H}_0$ as

$$\mathbf{H}_0 =$$

$$\begin{bmatrix} \ddots & \vdots & \vdots & \vdots & \vdots & \vdots & \vdots & \vdots & \\ & h_0(L-1) & h_0(L-2) & \cdots & \cdots & h_0(0) & 0 & 0 & \ddots \\ & 0 & 0 & h_0(L-1) & \cdots & h_0(2) & h_0(1) & h_0(0) & \\ & \vdots & \vdots & \vdots & \vdots & \vdots & \vdots & \vdots & \end{bmatrix}$$

(2.6)

which has as its kth row the elements of the sequence $h_0(2k - n)$. Premultiplying by $\mathbf{H}_0$ correpsonds to filtering by $H_0(z)$ followed by subsampling by a factor of 2. Also define

$$\mathbf{G}_0^T =$$

$$\begin{bmatrix} \ddots & \vdots & \vdots & \vdots & \vdots & \vdots & \vdots & \vdots & \\ & g_0(0) & g_0(1) & \cdots & \cdots & g_0(L-1) & 0 & 0 & \ddots \\ & 0 & 0 & g_0(0) & \cdots & g_0(L-3) & g_0(L-2) & g_0(L-1) & \\ & \vdots & \vdots & \vdots & \vdots & \vdots & \vdots & \vdots & \end{bmatrix},$$

(2.7)

so $\mathbf{G}_0$ has as its kth column the elements of the sequence $g_0(n - 2k)$. Premultiplying by $\mathbf{G}_0$ corresponds to upsampling by two followed by filtering by $G_0(z)$. Define $\mathbf{H}_1$ by replacing the coefficients of $h_0(n)$ with those of $h_1(n)$ in (2.6) and $\mathbf{G}_1$ by replacing the coefficients of $g_0(n)$ with those of $g_1(n)$ in (2.7).

We find that (2.4) gives that all rows of $\mathbf{H}_1$ are orthogonal to all columns of $\mathbf{G}_0$. Similarly we find, from (2.5), that all of the columns of $\mathbf{G}_1$ are orthogonal to the rows of $\mathbf{H}_0$. So in matrix notation:

$$(2.8) \qquad\qquad \mathbf{H}_0 \mathbf{G}_1 = 0 = \mathbf{H}_1 \mathbf{G}_0.$$

Now $P(z) = G_0(z)H_0(z) = 2 \cdot H_0(z)H_1(-z)/\Delta_m(z)$ and $P(-z) = G_1(z)$ $H_1(z)$ are both valid and have the form given in Figure 2.2 (a). Hence the impulse responses of $g_i(n)$ and $h_i(n)$ are orthogonal with respect to even shifts

$$(2.9) \qquad\qquad < g_i(n), h_i(2l - n) > \; = \; \delta_l.$$

In operator notation:

$$(2.10) \qquad\qquad \mathbf{H}_0 \mathbf{G}_0 = \mathbf{I} = \mathbf{H}_1 \mathbf{G}_1.$$

Since we have a perfect reconstruction system we get:

$$(2.11) \qquad\qquad \mathbf{G}_0 \mathbf{H}_0 + \mathbf{G}_1 \mathbf{H}_1 = \mathbf{I}.$$

Of course (2.11) indicates that no nonzero vector can lie in the column nullspaces of both $\mathbf{G}_0$ and $\mathbf{G}_1$. Note that (2.10) implies that $\mathbf{G}_0\mathbf{H}_0$ and $\mathbf{G}_1\mathbf{H}_1$ are each projections (since $\mathbf{G}_i\mathbf{H}_i\mathbf{G}_i\mathbf{H}_i = \mathbf{G}_i\mathbf{H}_i$). They project onto subspaces which are not in general orthogonal (since the operators are not self-adjoint). Because of (2.4), (2.5) and (2.9) the analysis/synthesis system is termed biorthogonal. If we interleave the rows of $\mathbf{H}_0$ and $\mathbf{H}_1$, and form a block-Toeplitz matrix

$$\mathbf{A} =
\begin{bmatrix}
 & \vdots & \vdots & & & \vdots & \vdots & \vdots & \vdots & \vdots & \\
 & h_0(L-1) & h_0(L-2) & \cdots & \cdots & h_0(0) & 0 & 0 & \\
\ddots & h_1(L-1) & h_1(L-2) & \cdots & \cdots & h_1(0) & 0 & 0 & \ddots \\
 & 0 & 0 & h_0(L-1) & \cdots & h_0(2) & h_0(1) & h_0(0) & \\
 & 0 & 0 & h_1(L-1) & \cdots & h_1(2) & h_1(1) & h_1(0) & \\
 & \vdots & \vdots & \vdots & \vdots & \vdots & \vdots & \vdots & \\
\end{bmatrix},$$

(2.12)
we find that the rows of $\mathbf{A}$ form a basis for $l^2(Z)$. If we form $\mathbf{B}$ by interleaving the columns of $\mathbf{G}_0$ and $\mathbf{G}_1$ we find

$$\mathbf{B} \cdot \mathbf{A} = \mathbf{I}.$$

In the special case where we have a unitary solution one finds: $\mathbf{G}_0 = \mathbf{H}_0^T$ and $\mathbf{G}_1 = \mathbf{H}_1^T$, and (2.8) gives that we have projections onto subspaces which are mutually orthogonal. The system then simplifies to the orthogonal case, where $\mathbf{B} = \mathbf{A}^{-1} = \mathbf{A}^T$.

A point that we wish to emphasize is that in the conditions for perfect reconstruction, (2.2) and (2.3), the filters $H_0(z)$ and $G_0(z)$ are related via their product $P(z)$. It is the choice of the function $P(z)$ and the factorization taken that determines the properties of the filter bank. We conclude the introduction with a lemma that sums up the foregoing [18].

LEMMA 2.1. *To design a two-channel perfect reconstruction filter bank it is necessary and sufficient to find a $P(z)$ satisfying (2.3), factor it $P(z) = G_0(z)H_0(z)$ and assign the filters as given in (2.2).*

2.1. Deriving continuous-time bases from discrete-time ones.
We have seen that the construction of bases from discrete-time signals can be easily accomplished by using a perfect reconstruction filter bank as the basic building block. This gives us bases that have a certain structure, and for which the analysis and synthesis can be efficiently performed. The design of bases for continuous-time signals appears more difficult. However, it works out that we can mimic many of the ideas used in the discrete-time case, when we go about the construction of continuous-time bases.

In fact, there is a very close correspondence between the discrete-time bases generated by two-channel filter banks, and dyadic wavelet bases. These are continuous-time bases formed by the stretches and translates of a single function, where the stretches are integer powers of two:

$$(2.13) \qquad \{\psi_{jk}(x) = 2^{-j/2}\psi(2^{-j}x - k), \quad j, k, \in Z\}$$

This relation has been thoroughly explored in [10], [48].

To be precise, a basis of the form in (2.13) necessarily implies the existence of an underlying two-channel filter bank. Conversely a two-channel filter bank can be used to generate a basis as in (2.13) provided that the lowpass filter $H_0(z)$ is *regular* (to be explained below). It is not our intention to go too deeply into the details of this connection, but a brief review of the generation of wavelets from filter banks follows. Interested readers might consult [10] or [48].

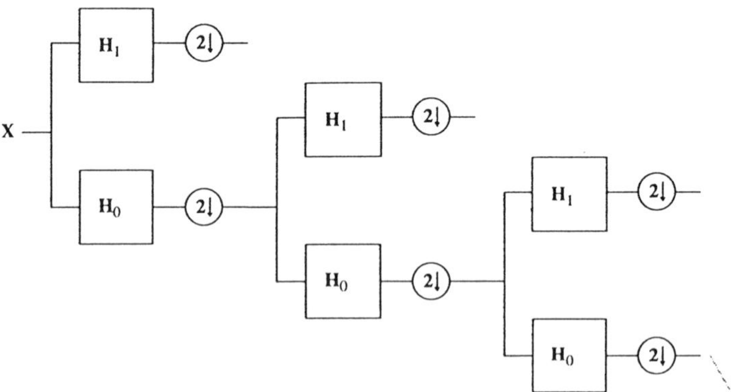

FIG. 2.3. *Iterated filter bank used for obtaining the continuous-time wavelet bases.*

Considering the octave-band tree of discrete-time filters in Figure 2.3, one notices that the lower branch is a cascade of filters $H_0(z)$ followed by subsampling by 2. It is easily shown [48], that the cascade of i blocks of filtering operations, followed by subsampling by 2, is equivalent to a filter $H_0^{(i)}(z)$ with z-transform:

$$(2.14) \qquad H_0^{(i)}(z) = \prod_{l=0}^{i-1} H_0(z^{2^l}), \qquad i = 1, 2 \cdots,$$

followed by subsampling by 2^i. We define $H_0^{(0)}(z) = 1$ to initialize the recursion. Now in addition to the discrete-time filter, consider the function $f^{(i)}(x)$ which is piecewise constant on intervals of length $1/2^i$, and equal to:

$$(2.15) \qquad f^{(i)}(x) = 2^{i/2} \cdot h_0^{(i)}(n), \qquad n/2^i \le x < (n+1)/2^i.$$

Note that the normalization by $2^{i/2}$ ensures that if $\sum (h_0^{(i)}(n))^2 = 1$ then $\int (f^{(i)}(x))^2 dx = 1$ as well. Also, it can be checked that $\|h_0^{(i)}\|_2 = 1$ when $\|h_0^{(i-1)}\|_2 = 1$. The relation between the sequence $H_0^{(i)}(z)$ and the function $f^{(i)}(x)$ is clarified in Figure 2.4, where the first three iterations of each is shown for the simple case of a filter of length 4. We are going to use the sequence of functions $f^{(i)}(x)$ to converge to the *scaling function* $\varphi(x)$ of a wavelet basis. Hence, a fundamental question is to find out whether and to what the function $f^{(i)}(x)$ converges as $i \to \infty$. First assume that the filter $H_0(z)$ has a zero at the half sampling frequency, or $H_0(e^{j\pi}) = 0$. This together with the fact that the filter impulse response is orthogonal to its even translates is equivalent to $\sum h_0(n) = H_0(1) = \sqrt{2}$. Define $M_0(z) = 1/\sqrt{2} \cdot H_0(z)$, that is $M_0(1) = 1$. Now factor $M_0(z)$ into its roots at π (we assume that there is at least one) and a remainder polynomial $K(z)$, in the following way:

$$M_0(z) = [(1 + z^{-1})/2]^N K(z).$$

Note that $K(1) = 1$ from the definitions. Now call B the supremum of $|K(z)|$ on the unit circle:

$$B = \sup_{\omega \in [0, 2\pi]} |K(e^{j\omega})|.$$

Then the following result from [10] holds:

PROPOSITION 2.2 (DAUBECHIES 1988). *If $B < 2^{N-1}$, and*

$$(2.16) \qquad \sum_{n=-\infty}^{\infty} |k(n)|^2 |n|^\epsilon \; < \; \infty, \; \textit{for some } \epsilon > 0,$$

then the piecewise constant function $f^{(i)}(x)$ defined in (2.15) converges pointwise to a continuous function $f^{(\infty)}(x)$.

This is a sufficient condition to ensure pointwise convergence to a continuous function, and can be used as a simple test. We shall refer to any filter for which the infinite product converges as *regular*.

If we indeed have convergence, then we define

$$f^{(\infty)}(x) = \varphi(x)$$

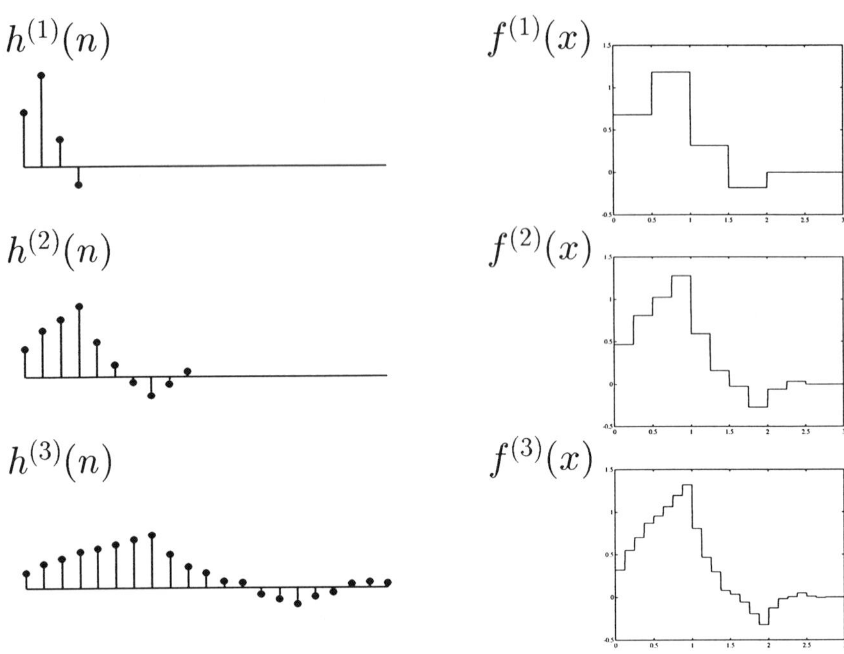

FIG. 2.4. *Iterations of the discrete-time filter (2.14) and the continuous-time function (2.15) for the case of a length-4 filter $H_0(z)$. The length of the filter $H_0^{(i)}(z)$ increases without bound, while the function $f^{(i)}(x)$ actually has bounded support from ([18]).*

as the analysis scaling function, and

$$(2.17) \qquad \psi(x) = 2^{-1/2} \sum h_1(n) \varphi(2x - n),$$

as the analysis wavelet. It can be shown, that if the filters $h_0(n)$ and $h_1(n)$ are from a perfect reconstruction filter bank, that (2.13) then indeed forms a continuous-time basis.

In a similar way we examine the cascade of i blocks of the synthesis filter $g_0(n)$

$$(2.18) \qquad G_0^{(i)}(z) = \prod_{l=0}^{i-1} G_0(z^{2^l}), \qquad i = 1, 2 \cdots.$$

Again, define $G_0^{(0)}(z) = 1$ to initialize the recursion, and normalize $G_0(1) = 1$. From this define a function which are piecewise constant on intervals of length $1/2^i$:

$$(2.19) \qquad \tilde{f}^{(i)}(x) = 2^{i/2} \cdot g_0^{(i)}(-n), \qquad n/2^i \le x < (n+1)/2^i.$$

We call the limit $\tilde{f}^{(\infty)}(x)$, if it exists, $\tilde{\varphi}(x)$ the synthesis scaling function, and we find

$$(2.20) \qquad \tilde{\psi}(x) = 2^{1/2} \cdot \sum_{n=0}^{L-1} g_1(-n) \cdot \tilde{\varphi}(2x - n).$$

The biorthogonality properties of the analysis and synthesis continuous-time functions follow from the corresponding properties of the discrete-time ones. That is (2.9) leads to

$$(2.21) \qquad < \tilde{\varphi}(x), \varphi(x - k) > = \delta_k,$$

and

$$(2.22) \qquad < \tilde{\psi}(x), \psi(x - k) > = \delta_k.$$

Similarly

$$(2.23) \qquad < \tilde{\varphi}(x), \psi(x - k) > = 0,$$

$$(2.24) \qquad < \tilde{\psi}(x), \varphi(x - k) > = 0,$$

come from (2.4) and (2.5) respectively.

We have shown that the conditions for perfect reconstruction on the filter coefficients lead to functions that have the biorthogonality properties as shown above. Orthogonality across scales is also easily verified:

$$< \tilde{\psi}(2^j x), \psi(2^i x - k) > = \delta_{i-j} \delta_k.$$

Thus the set $\{\psi(2^j x), \tilde{\psi}(2^i x - k), i, j, k \in Z\}$ is biorthogonal. That it is complete can be verified as in the orthogonal case [5]. Hence any function from $L^2(R)$ can be written:

$$f(x) = \sum_j \sum_l < f(x), 2^{-j/2}\psi(2^j x - l) > 2^{-j/2}\tilde{\psi}(2^j x - l).$$

Note that $\psi(x)$ and $\tilde{\psi}(x)$ play interchangeable roles.

2.2. Two-channel filter banks and wavelets. We have seen that the design of discrete-time bases is not difficult: using two-channel filter banks as the basic building block they can be easily derived. We also know that, using (2.15) and (2.19), we can generate continuous-time bases quite easily also. If we were just interested in the construction of bases, with no further requirements, we could stop here. However, for applications such as compression, we will often be interested in other properties of the basis functions. For example whether or not they have any symmetry, or finite support, and whether or not the basis is an orthonormal one. We examine these three questions for the remainder of this section.

From the filter bank point of view the properties we are most interested in are the following:
- **Orthogonality:**

$$(2.25)\quad < h_0(n), h_0(n + 2k) > \, = \, \delta_k \, = \, < h_1(n), h_1(n + 2k) >,$$

$$(2.26)\quad\quad\quad < h_0(n), h_1(n + 2k) > \, = \, 0.$$

- **Linear phase:** $H_0(z)$, $H_1(z)$, $G_0(z)$ and $G_1(z)$ are all linear phase filters.
- **Finite support:** $H_0(z)$, $H_1(z)$, $G_0(z)$ and $G_1(z)$ are all FIR filters.

The reason for our interest is twofold. Firstly, these properties are possibly of value in perfect reconstruction filter banks used in subband coding schemes. For example orthogonality implies that there will be energy conservation in the channels; linear phase is possibly of interest in very low bit-rate coding of images, and FIR filters have the advantage of having very simple low-complexity implementations. Secondly, these properties are carried over to the wavelets that are generated. So if we design a filter bank with a certain set of properties the continuous-time basis that it generates will also have these properties.

LEMMA 2.3. *If the filters belong to an orthogonal filter bank, we shall have*

$$< \varphi(x), \varphi(x + k) > \, = \, \delta_k \, = < \psi(x), \psi(x + k) >,$$

$$< \varphi(x), \psi(x + k) > \ = \ 0.$$

Proof: From the definition (2.15) $f^{(0)}(x)$ is just the indicator function on the interval $[0, 1)$; so we immediately get orthogonality at the 0th level, that is: $< f^{(0)}(x - l), f^{(0)}(x - k) > \ = \delta_{kl}$. Now we assume orthogonality at the ith level:

$$(2.27) \qquad < f^{(i)}(x - l), f^{(i)}(x - k) > \ = \ \delta_{kl},$$

and prove that this implies orthogonality at the $(i + 1)$st level:

$$< f^{(i+1)}(x - l), f^{(i+1)}(x - k) > \ =$$

$$= \ 2 \sum_n \sum_m h_0(n) h_0(m) < f^{(i)}(2x - 2l - n), f^{(i)}(2x - 2k - m) >,$$

$$= \ \sum_n h_0(n) h_0(n + 2l - 2k),$$

$$= \ \delta_{kl}.$$

Hence by induction (2.27) holds for all i. So in the limit $i \to \infty$:

$$(2.28) \qquad < \varphi(x - l), \varphi(x - k) > \ = \ \delta_{kl}.$$

Similarly for the other cases. $\qquad \qquad \square$

The orthogonal case gives considerable simplification, both in the discrete-time and continuous-time cases. Next consider the implication of using FIR filters.

LEMMA 2.4. *If the filters belong to an FIR filter bank then $\varphi(x)$, $\psi(x)$, $\tilde{\varphi}(x)$ and $\tilde{\psi}(x)$ will have support on some finite interval.*

Proof: The filter $H_0^{(i)}(z)$ and $G_0^{(i)}(z)$ defined in (2.14) have respective lengths $(2^i - 1)(L_a - 1) + 1$ and $(2^i - 1)(L_s - 1) + 1$ where L_a and L_s are the lengths of $H_0(z)$ and $G_0(z)$. Hence $f^{(i)}(x)$ in (2.15) is supported on the interval $[0, L_a - 1)$ and $\tilde{f}^{(i)}(x)$ on the interval $[0, L_s - 1)$. This holds $\forall \ i$; hence in the limit $i \to \infty$ this gives the support of the scaling functions $\varphi(x)$ and $\tilde{\varphi}(x)$. That $\psi(x)$ and $\tilde{\psi}(x)$ have bounded support follow from (2.17) and (2.20). $\qquad \qquad \square$

Finally linear phase filters imply symmetric or antisymmetric wavelets.

LEMMA 2.5. *If the filters belong to a linear phase filter bank then $\varphi(x)$, $\psi(x)$, $\tilde{\varphi}(x)$ and $\tilde{\psi}(x)$ will be symmetric or antisymmetric.*

Proof: The filter $H_0^{(i)}(z)$ will have linear phase if $H_0(z)$ does. If $H_0^{(i)}(z)$ has length $(2^i - 1)(L_a - 1) + 1$ the point of symmetry is $(2^i - 1)(L_a - 1)/2$

which need not be an integer. The point of symmetry for $f^{(i)}(x)$ will then be $[(2^i - 1)(L_a - 1) + 1]/2^{i+1}$ or $[(2^i - 1)(L_a - 1) + 2]/2^{i+1}$. In either case, by taking the limit $i \to \infty$ we find that $\varphi(x)$ is symmetric about the point $(L_a - 1)/2$. Similarly for the other cases. $\qquad\square$

Thus having established the relation between wavelets and filter banks we can examine the structure of filter banks in detail, and afterward use them to generate wavelets as described above. It should be emphasized that we are speaking of the two-channel one-dimensional case. The multi-dimensional case will be examined in Section 3.

2.3. Structure of two-channel filter banks. We saw already that it is the choice of the function $P(z)$ and the factorization taken that determines the properties of the filter bank. In terms of $P(z)$ we give necessary and sufficient conditions for the three properties mentioned above:

- **Orthogonality:** $P(z)$ is an autocorrelation; and $H_0(z)$ and $G_0(z)$ are its spectral factors.
- **Linear phase:** $P(z)$ is linear phase, and $H_0(z)$ and $G_0(z)$ are its linear phase factors.
- **Finite support:** $P(z)$ is FIR, and $H_0(z)$ and $G_0(z)$ are its FIR factors.

Obviously the factorization is not unique in any of the cases above. The orthogonal case has been examined in [21] [40] The FIR case has been examined in detail in [10], [42], [47], [48] and the linear phase case in [5], [35], [47], [48]. The proofs of the conditions given above for linear phase and finite support are obvious, and are omitted; the proof of the condition for orthogonality is included.

LEMMA 2.6. *To have an orthogonal filter bank it is necessary and sufficient that $P(z)$ be an autocorrelation, and that $H_0(z)$ and $G_0(z)$ be its spectral factors.*

Proof: Equation (2.26) implies that every second term of the cross-correlation between the filters $H_0(z)$ and $H_1(z)$ is zero

$$(2.29) \qquad H_0(z)H_1(z^{-1}) + H_0(-z)H_1(-z^{-1}) = 0.$$

This implies that

$$(2.30) \qquad H_1(z) = z^{-1}\alpha(z^2)H_0(-z^{-1}),$$

for some function $\alpha(z)$. Equally, (2.25) implies that every second term but one of the autocorrelations of $H_0(z)$ and $H_1(z)$ are zero

$$(2.31) \quad H_i(z)H_i(z^{-1}) + H_i(-z)H_i(-z^{-1}) = 2 \qquad i \epsilon \{0,1\}.$$

In order for this to be true for both $i = 0$ and $i = 1$ requires that $\alpha(z)\alpha(z^{-1}) = 1$, that is $\alpha(z)$ is an allpass function.

Now construct $P(z)$

$$
\begin{aligned}
P(z) &= H_0(z)H_1(-z)C(z) \\
&= \frac{-z^{-1}H_0(z)H_0(z^{-1})\alpha(z^2)}{-z^{-1}\alpha(z^2)[H_0(z)H_0(z^{-1}) + H_0(-z)H_0(-z^{-1})]} \\
&= H_0(z)H_0(z^{-1})
\end{aligned}
$$

which is always an autocorrelation. Since $G_0(z) = H_0(z^{-1})$ above it is clear that $H_0(z)$ and $G_0(z)$ are the spectral factors of $P(z)$. $\qquad\square$

Having seen that the design problem can be considered in terms of $P(z)$ and its factorizations, we consider the three conditions of interest, from this point of view.

2.3.1. Orthogonal solutions. In the case where the filter bank is to be orthogonal we can get a complete constructive characterization of the solutions, as given by the following theorem, taken from [21].

THEOREM 2.7. *All orthogonal rational two channel filter banks can be formed as follows: (i) Choosing an arbitrary polynomial $R(z)$, form:*

$$
P(z) = \frac{2 \cdot R(z)R(z^{-1})}{R(z)R(z^{-1}) + R(-z)R(-z^{-1})},
$$

(ii) Factor as $P(z) = H(z)H(z^{-1})$,
(iii) Form the filter $H_0(z) = A_0(z)H(z)$, where $A_0(z)$ is an arbitrary all-pass,
(iv) Choose $H_1(z) = z^{2k-1}H_0(-z^{-1})A_1(z^2)$, where $A_1(z)$ is again an arbitrary allpass.
(v) Choose $G_0(z) = H_0(z^{-1})$, and $G_1(z) = -H_1(z^{-1})$.

For a proof see [21], [18].

Example: Take $R(z) = (1 + z^{-1})^N$ as above and $N = 7$. It works out that in this case there is a closed form factorization for the filters [21].

$$
\begin{aligned}
P(z) &= \frac{(1,14,91,364,1001,2002,3003,3432,3003,2002,1001,364,91,14,1)\cdot z^7}{14z^6+364z^4+2002z^2+3432+2002z^{-2}+364z^{-4}+14z^{-6}} \\
&= \frac{E(z)E(z^{-1})}{K(z)K(z^{-1})},
\end{aligned}
$$

where

$$
\frac{E(z)}{K(z} = \frac{(1 + 7z^{-1} + 21z^{-2} + 35z^{-3} + 35z^{-4} + 21z^{-5} + 7z^{-6} + z^{-7})}{\sqrt{2}\cdot(1 + 21z^{-2} + 35z^{-4} + 7z^{-6})}.
$$

Note that we have used the following shorthand notation to list the coefficients of a causal FIR sequence:

$$
\sum_{n=0}^{N-1} a_n z^{-n} = (a_0, a_1, a_2, \cdots a_{N-1}).
$$

So using the description of the filters in Theorem 2.7, with the simplest case $A_0(z) = A_1(z) = 1$ and $k = 0$ we find:

$$H_0(z) \;=\; \frac{(1 + 7z^{-1} + 21z^{-2} + 35z^{-3} + 35z^{-4} + 21z^{-5} + 7z^{-6} + z^{-7})}{\sqrt{2} \cdot (1 + 21z^{-2} + 35z^{-4} + 7z^{-6})}$$

$$H_1(z) \;=\; z^{-1}\frac{(1 - 7z^{1} + 21z^{2} - 35z^{3} + 35z^{4} - 21z^{5} + 7z^{6} - z^{7})}{\sqrt{2} \cdot (1 + 21z^{2} + 35z^{4} + 7z^{6})}$$

$$G_0(z) \;=\; H_0(z^{-1}) \qquad G_1(z) = H_1(z^{-1}).$$

In the notation of Proposition 2.2, $B = 8 < 2^6$ so that for this choice of $H_0(z)$ the left-hand side of (2.15) converges to a continuous function. The wavelet, scaling function and their spectra are shown in Figure 2.5. This is in fact a seventh order halfband Butterworth filter.

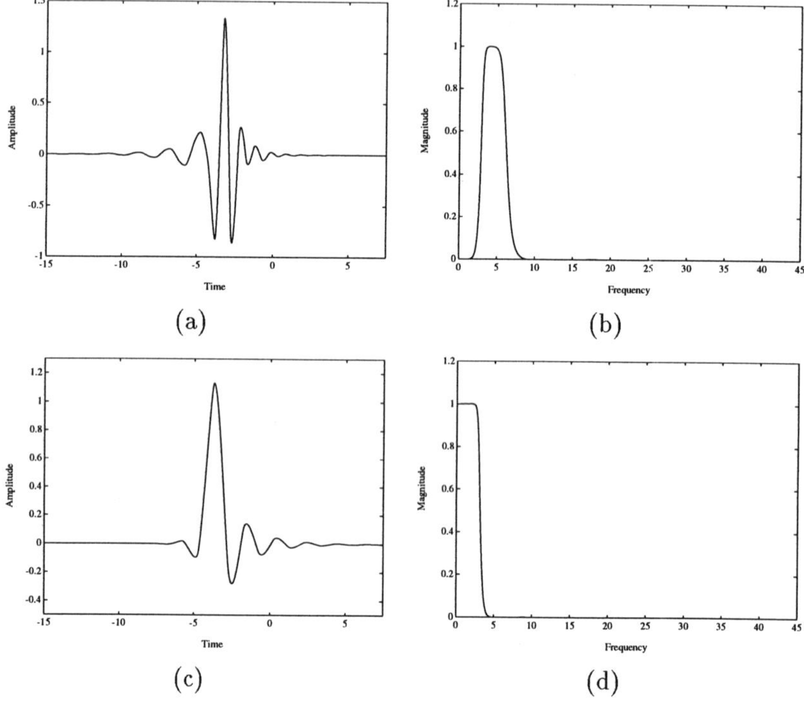

FIG. 2.5. *Example of Butterworth orthogonal wavelet; here $N = 7$, and the closed form factorization has been used. (a) The wavelet. (b) Spectrum of the wavelet. (c) Scaling function. (d) Spectrum of the scaling function.*

2.3.2. Finite impulse response and symmetric solutions. In the case where the filters are to be FIR we merely require that $P(z)$ be FIR; it is trivially easy to design one. Similarly to have symmetric filters, we merely force $P(z)$ to be symmetric. Obviously any symmetric $P(z)$, which is FIR and satisfies (2.3) can be used to give symmetric FIR filters. We would like in addition that the lowpass filters are regular, so that we get symmetric bounded support continuous-time basis functions.

One strategy would be to design a $P(z)$ with the desired properties and then factor to find the filters. Alternatively, we can choose one of the factors, and then find the other necessary to make the product $P(z)$ satisfy (2.3). We will use this approach and, to ensure regularity, choose one factor to be $(1 + z^{-1})^{2N}$ This can be done by solving a linear system of equations [48].

Example If we choose $N = 3$ we must find the complement to $(1 + z^{-1})^6$; so we solve the 3 by 3 system found by imposing the constraints on the coefficients of the odd powers of z^{-1} of

$$P(z) = (k_0 + k_1 z^{-1} + k_2 z^{-2} + k_1 z^{-3} + k_0 z^{-4})$$

$$\cdot (1 + 6z^{-1} + 15z^{-2} + 20z^{-3} + 15z^{-4} + 6z^{-5} + z^{-6}) \cdot z^5.$$

Multiplying out, we find that the condition on the k_i to ensure that $P(z)$ satisfies (2.3) can be expressed

$$\begin{pmatrix} 6 & 1 & 0 \\ 20 & 16 & 6 \\ 12 & 30 & 20 \end{pmatrix} \begin{pmatrix} k_0 \\ k_1 \\ k_2 \end{pmatrix} = \begin{pmatrix} 0 \\ 0 \\ 1 \end{pmatrix},$$

giving $\mathbf{k}_6 = (3/2, -9, 19)/128$. So choosing $H_0(z) = (1 + z^{-1})^6$ and $G_0(z) = K_6(z)$ gives one linear phase FIR solution, but many other factorizations are possible.

In general, if $P(z)$ is to have a factor $(1 + z^{-1})^{2N}$, we solve the system:

$$(2.32) \qquad\qquad \mathbf{F}_{2N} \cdot \mathbf{k}_{2N} = \mathbf{e}_{2N},$$

where $\mathbf{F}_{2N}$ is the $N \times N$ matrix, $\mathbf{k}_{2N} = (k_0, \cdots, k_{(N-1)})$ and $\mathbf{e}_{2N}$ is the length N vector $(0, 0, \cdots, 1)$.

Having found the coefficients of $K_{2N}(z)$ we factor it into linear phase components; and then regroup these factors of $K_{2N}(z)$, and the $2N$ zeros at $z = -1$ to form two filters: $H_0(z)$ and $G_0(z)$, both of which are to be regular. An example where $N = 9$ is given in Figure 2.6.

2.4. Summary of two-channel solutions. An important consideration that is often encountered in the design of wavelets, or of the filter banks that generate them, is the necessity of satisfying competing design constraints. This makes it necessary to clearly understand whether desired properties are mutually exclusive.

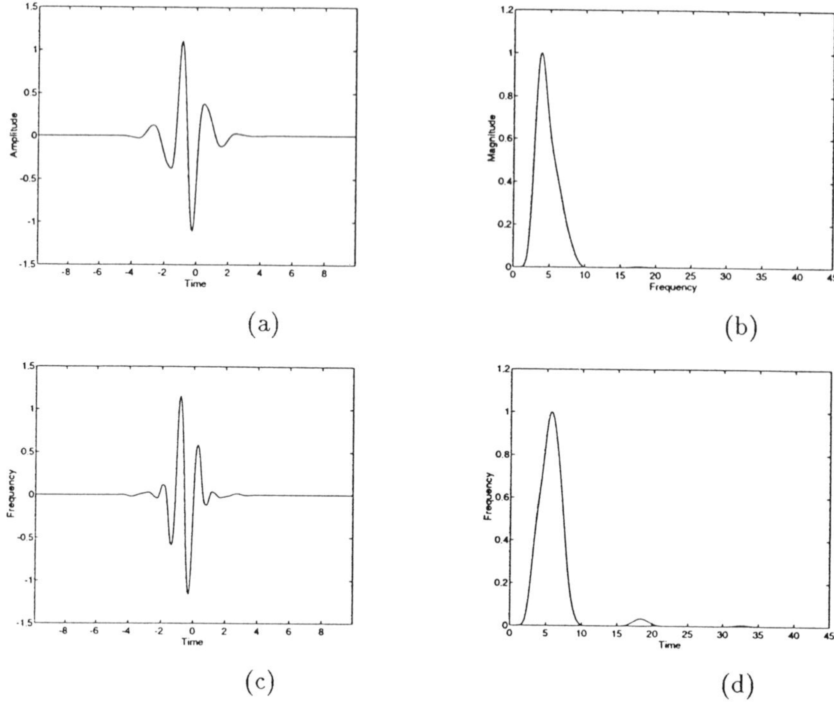

FIG. 2.6. *Biorthogonal wavelets generated by filters of length 18 given in [48]. (a) Analysis wavelet function $\psi(x)$. (b) Spectrum of analysis wavelet. (c) Synthesis wavelet function $\tilde{\psi}(x)$. (d) Spectrum of synthesis wavelet.*

Perfect reconstruction solutions, with the constraint that $P(z)$ be rational with real coefficients, must satisfy (2.3). Such general solutions, which do not necessarily have additional properties were given in [42].

A Venn diagram of the designs we have considered is shown in Figure 2.7. The solutions of set A, where all of the filters involved are FIR, were studied in [42], [47]. Set B contains all orthogonal solutions, and was the main focus of [21] [38] [40]. A complete characterization of this set was given in Theorem 2.7. A very different characterization, based on lattice structures is given in [14]. A particular subset of these set permits a closed form spectral factorization of the $P(z)$ function [21]. Set C contains the solutions where all filters are linear phase, first examined in [47].

The earliest examples of perfect reconstruction solutions [32], [41] were orthogonal and FIR; *i.e.* they were in $A \cap B$. A constructive parametrization of $A \cap B$ was given in [44]. The construction and characterization of examples which converge to wavelets was first done in [10]. Filter banks with FIR linear phase filters (*i.e.* $A \cap C$) were first given in [47], and also studied in terms of lattices in [35], [50]. The construction of wavelet examples is given in [5] and [48]. Filter banks which are linear phase and orthogonal were first presented in [21].

That there exist only trivial solutions which are linear phase, orthogonal and FIR is indicated by the intersection $A \cap B \cap C$; the only solutions are two tap filters [10], [45], [48].

It warrants emphasis that Figure 2.7 illustrates the filter bank solutions; if the filters are regular then they will lead to wavelets. Of the dyadic wavelet bases known to the authors the only ones based on filters where $P(z)$ is not rational are those of Meyer [31], and the only ones where the filter coefficients are complex are those of Lawton [25]. For the case of the Battle-Lemarié wavelets [2], [27], while the filters themselves are not rational, the $P(z)$ function is; hence the filters would belong to $B \cap C$ in the figure.

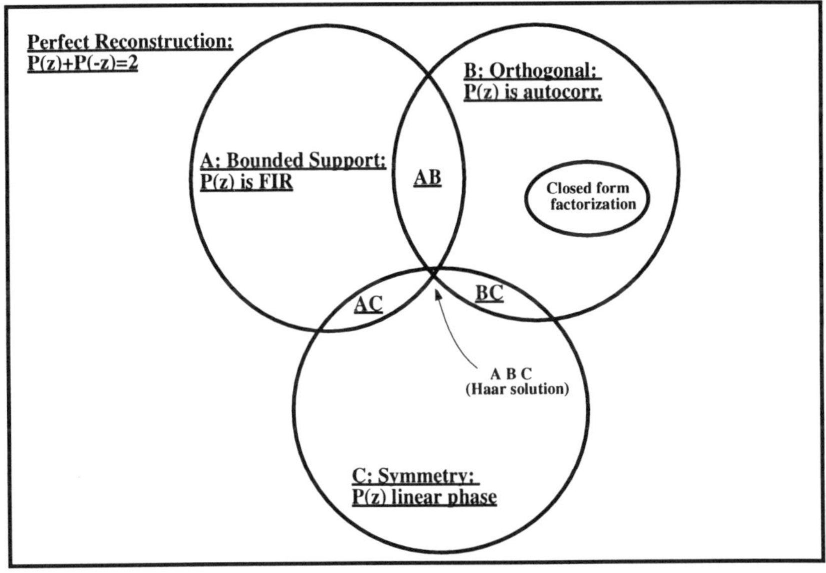

FIG. 2.7. *Two channel perfect reconstruction filter banks. The Venn diagram illustrates which competing constraints can be simultaneously satisfied. The sets A, B, C contain FIR, orthogonal and linear phase solutions respectively. Solutions in the intersection $A \cap B$ are examined in [42], [32], [44], [10]; those in the intersection $A \cap C$ are detailed in [47], [50], [35], [48], [5]; solutions in $B \cap C$ are constructed in [21]. The intersection $A \cap B \cap C$ contains only trivial solutions.*

3. Multidimensional filter banks and wavelets. The first question one would like to answer here is: why go to multiple dimensions? An obvious answer is that the whole area of filter banks has been heavily driven by applications, one of the recent ones being image compression. The way the problem was approached was to apply all known one-dimensional techniques separately, along one dimension at a time. Although a very simple solution, it suffers from some drawbacks: First, only separable, (*e.g.*, two-dimensional) filters are obtained in this way, leading to a fairly constrained design problem (nonseparable filters would offer $N_1 \cdot N_2$ free variables versus $N_1 + N_2$ in the separable case). Then, only rectangular divisions of the spectrum are possible (see Figure 3.1), while one might need divisions that would capture better the signal's energy concentration.

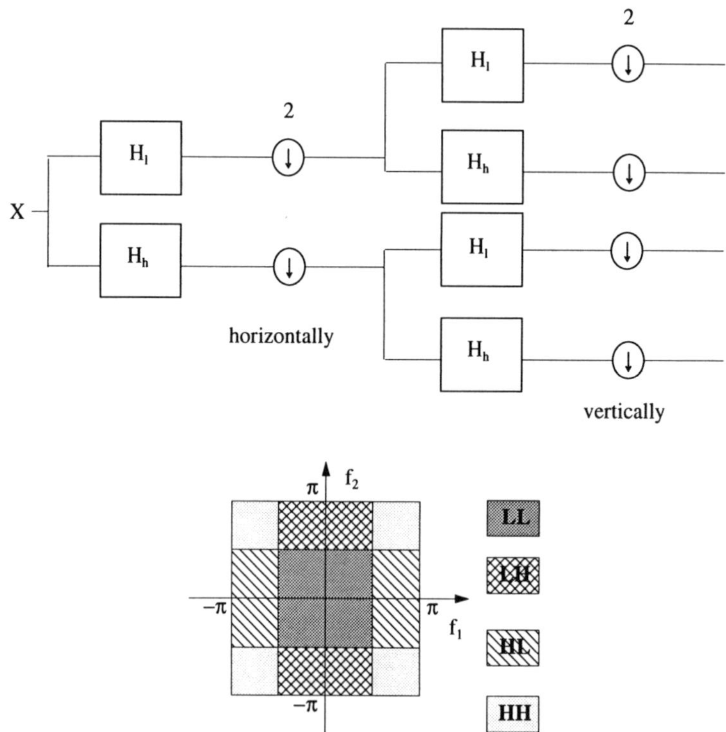

FIG. 3.1. *Scheme typically used for image processing, leading to a rectangular division of the spectrum.*

The first step toward using multidimensional techniques on multidimensional signals is to use the same kind of sampling as before (in the case of an image, sample first along the horizontal and then along the vertical dimensions), but use nonseparable filters instead of separable ones. It is obvious that one-dimensional techniques cannot be used anymore, although the sampling is separable. More sophisticated techniques use nonseparable sampling, which is represented by lattices[1]. Thus, true multidimensional treatment allows for nonseparable sampling, the advantage of which, as we have already said, is the possibility of obtaining nonrectangular divisions of the spectrum. The disadvantage, however, is that the analysis becomes more complex.

In trying to build multidimensional wavelet bases, one would encounter the same type of problems. An easy way to construct two-dimensional wavelets, for example, is to use tensor products of their one-dimensional counterparts, resulting, as will be seen later, in one scaling function and three different "mother" wavelets. Since now, scale change is represented by matrices, scaling matrix in this case will be $2\mathbf{I}$, that is, each dimension is dilated by 2. As for multidimensional filter banks, true multidimensional treatment of wavelets offers several advantages. First, one can still have a diagonal dilation (scaling) matrix, and yet design nonseparable (irreducible) scaling function and wavelets. Then, the scale change of $\sqrt{2}$, for example, is possible, leading to one scaling function and one wavelet, or a true two-dimensional counterpart of the well-known one-dimensional dyadic case. However, unlike for the filter banks, matrices used for dilation are more restricted, in that one requires dilation in each dimension. As in one dimension, the powerful connection with filter banks, through the method of iteration, can be exploited to design multidimensional wavelets. However, the task is much more involved due to incomplete cascade structures and the difficulty of imposing a zero of a particular order at aliasing frequencies. Regularity is much harder to achieve, and up to date, orthonormal families with arbitrarily high regularity, have not been found. A promising direction, however, has been the attempt to find transformations from one-dimensional into multidimensional filter banks, that would preserve perfect reconstruction and order of a zero. For more details on this topic, the reader is referred to [4], [24], [39].

3.1. Separable case. Separable sampling is represented by a diagonal sampling matrix denoting sampling along horizontal and vertical dimensions (in case of images). Take, for example, the matrix denoting subsampling by 2 in each dimension (the most commonly used one in image processing)

$$(3.1) \qquad \mathbf{D} = \begin{bmatrix} 2 & 0 \\ 0 & 2 \end{bmatrix}.$$

If one uses the scheme as in Figure 3.1 then all one-dimensional results are trivially extended to two dimensions. However, all limitations appearing

---

[1] A lattice is the set of all linear combinations of n basis vectors $\mathbf{a}_1, \mathbf{a}_2, \ldots, \mathbf{a}_n$, with integer coefficients, *i.e.*, a lattice is the set of all vectors generated by $\mathbf{Dk}$, $\mathbf{k} \in \mathcal{Z}^n$, where $\mathbf{D}$ is the matrix characterizing the sampling process.

in one dimension, will appear in two as well. For example, we know that, except for the Haar filters, there are no two-channel perfect reconstruction filter banks, being orthogonal and linear phase at the same time (in the FIR case). This implies that the same will hold in two dimensions if separable filters are used.

However, one could still sample separately and yet use nonseparable filters. In other words, one could have a direct 4-channel implementation of Figure 3.1 where the four filters could be H_0, H_1, H_2, H_3. While before $H_0(z_1, z_2) = H_0^{(1)}(z_1)H_0^{(2)}(z_2)$ where $H_0^{(1)}$ and $H_0^{(2)}$ are one-dimensional filters in horizontal and vertical dimensions, respectively, $H_0(z_1, z_2)$ is a true two-dimensional filter. The problem, however, becomes analysis and design of such a system.

For example, one can (similarly to the one-dimensional case), define polyphase components of filters, as its impulse responses on lattice points of a particular coset. For example, for a filter $f(n_1, n_2)$, the four polyphase components would be $f_{i,j}(n_1, n_2) = f(2n_1 + i, 2n_2 + j)$, for $i, j = 0, 1$. Then, the output of an analysis/synthesis system can be written as (see also (2.1))

$$(3.2) \quad Y = [z_1^{-1}\ z_2^{-1}\ z_2^{-1}\ z_1^{-1}\ 1]\ \mathbf{G}_p(z_1^2, z_2^2)\ \mathbf{H}_p(z_1^2, z_2^2)\ \mathbf{X}_p(z_1^2, z_2^2),$$

where $\mathbf{H}_p$ and $\mathbf{G}_p$ are analysis and synthesis polyphase matrices, respectively. The output of the system could also be written in terms of modulated versions of the filters $(H_k^{(i,j)} = H_k((-1)^i z_1, (-1)^j z_2))$, as a generalization of (2.1).

$$(3.3) \quad Y = \frac{1}{4}\mathbf{G} \cdot \begin{bmatrix} H_0^{(0,0)} & H_0^{(1,0)} & H_0^{(0,1)} & H_0^{(1,1)} \\ H_1^{(0,0)} & H_1^{(1,0)} & H_1^{(0,1)} & H_1^{(1,1)} \\ H_2^{(0,0)} & H_2^{(1,0)} & H_2^{(0,1)} & H_2^{(1,1)} \\ H_3^{(0,0)} & H_3^{(1,0)} & H_3^{(0,1)} & H_3^{(1,1)} \end{bmatrix} \cdot \begin{bmatrix} X^{(0,0)} \\ X^{(1,0)} \\ X^{(0,1)} \\ X^{(1,1)} \end{bmatrix},$$

where $\mathbf{G}$ is a row-vector containing synthesis filters, and $X^{(i,j)}$ is equal to $X((-1)^i z_1, (-1)^j z_2)$ (similarly for filters). Now, the condition for perfect reconstruction with FIR filters states that it is possible if and only if the determinant of the analysis polyphase matrix is a monomial, *i.e.*, $\det(\mathbf{H}_p(z_1, z_2)) = z_1^{-k_1} z_2^{-k_2}$.

One easy way to obtain such a determinant is by using cascade structures for building the polyphase matrix since one can easily impose perfect reconstruction together with some other desirable properties. For example, a cascade structure generating perfect reconstruction filters being both orthogonal and linear phase (not possible when using separable systems) is as follows:

$$(3.4) \quad \mathbf{H}_p(z_1, z_2) = \mathbf{R}_0 \prod_{i=1}^{k} \mathbf{D}(z_1, z_2)\mathbf{R}_i,$$

where $\mathbf{D}$ is the matrix of delays containing $(1, z_1^{-1}, z_2^{-1}, z_1^{-1}z_2^{-1})$ along the diagonal, and $\mathbf{R}_i$ are scalar persymmetric $(\mathbf{R} = \mathbf{JRJ})$ and unitary matrices.

3.2. Quincunx case. When the lattice we are using for sampling a multidimensional signal becomes nonseparable, design of perfect reconstruction filter banks becomesmore difficult.

We will examine the nonseparable case using quincunx sampling as an example, the reason being that it is the simplest multidimensional nonseparable lattice. Moreover, it subsamples by 2, *i.e.*, it is the real counterpart of the one-dimensional two-channel case we discussed in Section 2. One possible representation of the subsampling matrix is

$$(3.5) \qquad \mathbf{D}_Q \; = \; \begin{bmatrix} 1 & 1 \\ 1 & -1 \end{bmatrix}.$$

The input/output relationship is then

$$(3.6) \; Y = \frac{1}{2}[G_0 \; G_1] \begin{bmatrix} H_0(z_1, z_2) & H_0(-z_1, -z_2) \\ H_1(z_1, z_2) & H_1(-z_1, -z_2) \end{bmatrix} \begin{bmatrix} X(z_1, z_2) \\ X(-z_1, -z_2) \end{bmatrix}.$$

Similarly to the one-dimensional case, the orthogonality of the system is achieved when the following is satisfied:

$$(3.7) \;\; H_0(z_1, z_2)H_0(z_1^{-1}, z_2^{-1}) + H_0(-z_1, -z_2)H_0(-z_1^{-1}, -z_2^{-1}) \; = \; 2,$$

that is, the lowpass filter is orthogonal to its shifts on the quincunx lattice. Then, the highpass filter is of the following form:

$$(3.8) \qquad H_1(z_1, z_2) \; = \; -z_1^{-1}H_0(z_1^{-1}, z_2^{-1}),$$

and the synthesis filters are the same (within shift-reversal).

The design of such nonseparable systems is nontrivial. However, there are a few methods at our disposition. The first one is to use cascade structures, in a similar fashion as in the separable case

$$(3.9) \qquad \mathbf{H}_p(z_1, z_2) \; = \; \mathbf{R}_0 \cdot \prod_{j=1}^{k} \begin{bmatrix} 1 & 0 \\ 0 & z_1^{-1} \end{bmatrix} \mathbf{R}_{1_j} \begin{bmatrix} 1 & 0 \\ 0 & z_2^{-1} \end{bmatrix} \mathbf{R}_{2_j},$$

where $\mathbf{R}_{i_j}$ are scalar unitary matrices. A few other cascade structures can be found in [24], some of which produce filters which are linear phase. Another method is to use one to multidimensional transformations, that is, transforms where one could use one-dimensional perfect reconstruction systems and produce multidimensional ones preserving perfect reconstruction and other properties (*e.g.*, orthogonality and linear phase). One such transform is due to Ansari [1] and is based on separable polyphase components, but works only with IIR filters. Another one is the McClellan transformation [30] that preserves reconstruction but works only for linear phase filters. Since both methods are useful for constructing wavelets, the details of the constructions are given in the next section.

3.3. Wavelets in multiple dimensions. As we said earlier, scaling is now represented by a matrix $\mathbf{D}$. This matrix has to be well-behaved, that is,

1. $\mathbf{D}\mathcal{Z}^m \subset \mathcal{Z}^m$
2. $\mid \lambda_i \mid > 1$, for all i,

where λ_i are the eigenvalues of $\mathbf{D}$ [17]. The first condition requires $\mathbf{D}$ to have integer entries, while the second one states that all the eigenvalues of $\mathbf{D}$ must be strictly greater than 1, in order to ensure dilation in each dimension. For example, in the quincunx case, (3.5) and

$$(3.10) \qquad \mathbf{D}_{Q_2} = \begin{bmatrix} 1 & -1 \\ 1 & 1 \end{bmatrix},$$

are both valid matrices, while

$$(3.11) \qquad \mathbf{D}_{Q_3} = \begin{bmatrix} 2 & 1 \\ 0 & 1 \end{bmatrix},$$

is not, since it dilates only one dimension. The matrix $\mathbf{D}_Q$ from (3.5) is called a "symmetry" dilation matrix, used in [24], while $\mathbf{D}_{Q_2}$ is termed a "rotation" matrix used in [4]. Although both of these matrices represent the same lattice, they are fundamentally different when it comes to constructing wavelets.

For the case obtained as a tensor product, the dilation matrix is diagonal. Specifically, in two dimensions $\mathbf{D}_S$ is given in (3.1). The number of wavelets is determined by the number of cosets of $\mathbf{D}\mathcal{Z}^n$, $N-1$ or $\mid \det \mathbf{D} \mid - 1$. Thus, in the quincunx case, we have one "mother" wavelet, while in the separable case, there are three "mother" wavelets ψ_1, ψ_2, ψ_3.

The two-scale equation is obtained as in the one-dimensional case. For example, using $\mathbf{D}_Q$

$$(3.12)\ \varphi(x_1, x_2) = \frac{1}{\sqrt{2}} \sum_{n_1, n_2 \in \mathcal{Z}} h_0(n_1, n_2)\varphi(x_1 + x_2 - n_1, x_1 - x_2 - n_2).$$

3.3.1. Construction of wavelets using iterated filter banks. In what follows, we will concentrate on the quincunx case. Consider again Figure 2.3 with matrix $\mathbf{D}_Q$ replacing sampling by 2. Then the equivalent low branch after i steps of filtering and sampling by $\mathbf{D}_Q$ will be

$$(3.13) \qquad H^{(i)}(\omega_1, \omega_2) = \prod_{k=0}^{l-1} H_0\left((\mathbf{D}_Q^t)^i \begin{pmatrix} \omega_1 \\ \omega_2 \end{pmatrix} \right),$$

where $H^{(0)}(\omega_1, \omega_2) = 1$. Observe here that instead of scalar powers, we are dealing with taking powers of matrices. Thus, for different matrices, iterated filters are going to exhibit vastly different behavior, the most striking examples of which lead to fractal functions tiling the plane, and were independently discovered by Gröchenig and Madych [17] and Lawton and Resnikoff [26].

Now, as in the one-dimensional case, construct a continuous-time "graphical" function based on the iterated filter $h^{(i)}(n_1, n_2)$

$$(3.14) \qquad f^{(i)}(x_1, x_2) = 2^{\frac{i}{2}} h^{(i)}(n_1, n_2),$$

$$\begin{pmatrix} 1 & 1 \\ 1 & -1 \end{pmatrix}^i \begin{pmatrix} x_1 \\ x_2 \end{pmatrix} \in \begin{pmatrix} n_1 \\ n_2 \end{pmatrix} + [0, 1) \times [0, 1).$$

Note that these regions are not in general rectangular, and specifically in this case, they are squares in even, and diamonds (tilted squares) in odd iterations. Note that one of the advantages of using the matrix $\mathbf{D}_Q$ rather than $\mathbf{D}_{Q_2}$, is that it leads to separable sampling (diagonal matrix) in every other iteration, since $\mathbf{D}_Q^2 = 2\mathbf{I}$. The reason why this feature is useful is that one can use one-dimensional results in a separate manner in even iterations. We are again interested in the limiting behavior of this "graphical" function. Let us first assume that the limit of $f^{(i)}(x_1, x_2)$ exists and is in L^2, and we will come back later to the conditions under which it exists. Hence, we define the scaling function as

$$(3.15) \qquad \varphi(x_1, x_2) = \lim_{i \to \infty} f^{(i)}(x_1, x_2), \qquad \varphi(x_1, x_2) \in L^2(\mathcal{R}^2).$$

Once the scaling function exists, the wavelet can be obtained from

$$(3.16)\ \psi(x_1, x_2) = \frac{1}{\sqrt{2}} \sum_{n_1, n_2 \in \mathcal{Z}} h_1(n_1, n_2)\varphi(x_1 + x_2 - n_1, x_1 - x_2 - n_2).$$

Again, the coefficients used in (3.12) and (3.16) are the impulse response coefficients of the lowpass and highpass filters, respectively. To prove that the wavelet obtained in such a fashion actually produces an orthonormal basis for $L^2(\mathcal{R}^2)$, one has to demonstrate various facts. In the following statements the proofs are analogous to the one-dimensional ones, and can be found in [24]:

1. $< \varphi(\mathbf{D}_Q^i \mathbf{x} - \mathbf{n}), \varphi(\mathbf{D}_Q^i \mathbf{x} - \mathbf{k}) >= 2^{-i}\delta(\mathbf{n} - \mathbf{k})$, that is, the scaling function is orthogonal to its integer translates across all scales,
2. the same holds for the wavelet,
3. $< \varphi(\mathbf{x}), \psi(\mathbf{x} - \mathbf{k}) >$, the scaling function is orthogonal to the wavelet and its integer translates,
4. wavelets are orthogonal across scales.

These facts then state that the set

$$S = \{2^{-\frac{i}{2}}\psi(\mathbf{D}_Q^{-i}\mathbf{x} - \mathbf{n}) \mid i \in \mathcal{Z}, \mathbf{n} \in \mathcal{Z}^2, \mathbf{x} \in \mathcal{R}^2\},$$

is an orthonormal set. The only thing left to do is to show that the members of the set S constitute an orthonormal basis for $L^2(\mathcal{R}^2)$. The way to do it is to verify that S is a tight frame with a frame bound equal to one. The proof is omitted here, and is given in [24].

Another important fact is the necessity of at least one zero of the lowpass filter at aliasing frequencies. It holds in general, but will be given here for the quincunx case (the proof can be found in [24]).

PROPOSITION 3.1. *If the scaling function $\varphi(x_1, x_2)$ exists for some $(x_1, x_2) \in \mathcal{R}^2$, then*

$$(3.17) \qquad \sum_{\mathbf{k} \in \mathcal{Z}^2} h_0(\mathbf{D}_Q \mathbf{k} + \mathbf{k}_i) = \frac{1}{\sqrt{2}}, \qquad \mathbf{k}_0 \; = \; \begin{pmatrix} 0 \\ 0 \end{pmatrix}, \; \mathbf{k}_1 \; = \; \begin{pmatrix} 1 \\ 0 \end{pmatrix},$$

or, in other words

$$H_0(0,0) = \sqrt{2}, \quad H_0(\pi, \pi) = 0.$$

3.3.2. Design of multidimensional wavelets. As we have seen previously, the design of multidimensional filter banks is a difficult task from a signal processing point of view, but it becomes all the more involved by introducing the requirement that the lowpass filter be regular. Here, known techniques will be briefly reviewed, for more details the reader is referred to [4], [24].

Direct Design

To achieve perfect reconstruction in a subband system, cascade structures are perfect candidates, since beside perfect reconstruction, some other properties such as orthogonality and linear phase can be easily imposed. Recall that in one dimension, a zero of a sufficiently high order at π would guarantee the desired degree of regularity. Unfortunately, imposing a zero of a particular order in multiple dimensions becomes a nontrivial problem, and thus, algebraic solutions can be obtained just for very small size filters.

As an example of direct design, consider again the quincunx case with matrix $\mathbf{D}_Q$ and the cascade given in (3.9). Thus, the approach is to impose a zero of the highest possible order at (π, π) on the lowpass filter from (3.9)

$$(3.18) \qquad \frac{\partial^{k-1} H(\omega_1, \omega_2)}{\partial^l \omega_1 \partial^{k-l-1} \omega_2} \bigg|_{(\pi, \pi)} = 0, \qquad \begin{array}{l} k = 1, \ldots, m, \\ l = 0, \ldots, k-1. \end{array}$$

Upon imposing a second-order zero the following solutions are obtained

$$(3.19) \qquad a_0 = \pm\sqrt{3}, \quad a_1 = \pm\sqrt{3}, \quad a_2 = 2 \pm \sqrt{3},$$

$$(3.20) \qquad a_0 = \pm\sqrt{3}, \quad a_1 = 0, \quad a_2 = 2 \pm \sqrt{3}.$$

The solution in (3.20) is the one-dimensional D_2 filter [10], while (3.19) would be the smallest "regular" two-dimensional filter (actually, a counterpart of D_2). Figure 3.2 shows the tenth iteration of this solution. It was proven in [51] that this iterated function tends to a continuous function.

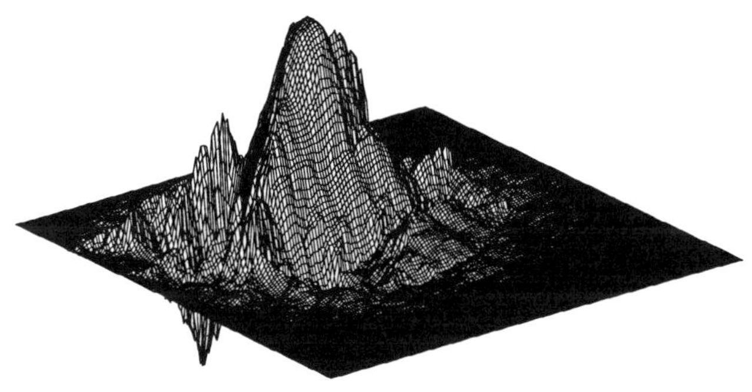

FIG. 3.2. *The tenth iteration of the smallest regular two-dimensional filter leading to the scaling function.*

This method, however, fails for larger size filters, since imposing a zero of a particular order means solving a large system of nonlinear equations (in the orthogonal case). Note, however, that numerical approaches are always possible [23].

One to Multidimensional Transformations

Another way to approach the design problem is to use transformations of one-dimensional filters into multidimensional ones in such a way that [24]

1. perfect reconstruction is preserved (in order to have a valid subband coding system) and
2. zeros at aliasing frequencies are preserved (necessary but not sufficient for regularity).

We will discuss two approaches. The first method involves using filters with separable polyphase components (see, for example, [1]). The second one is to use the McClellan transformation [30].

(i) Separable Polyphase Components

In this approach, the polyphase components of a multidimensional filter are obtained from the polyphase components of a prototype one-dimensional filter. We will concentrate on the two-dimensional case, but the same analysis can be carried out in more than two dimensions. Thus each polyphase component can be expressed as

$$H_i(z_1, z_2) \;=\; H_i(z_1)H_i(z_2).$$

The advantage of this method is that the implementation is simple (due to separable polyphase components) as well as the fact that the zeros at aliasing frequencies carry over. The problem with this approach, however, is that the perfect reconstruction property is preserved only for filters with allpass polyphase components, that is, IIR filters.

(ii) McClellan Transformation [30]

This transformation is well suited for the design of multidimensional filters since it leads to very efficient implementation. It transforms one-dimensional zero-phase filters into multidimensional zero-phase filters. Recently, the McClellan transformation has been recognized as a way of building multidimensional [39] as well as regular filter banks [4], [24]. To be more specific, if a filter is linear phase, then it can be written as (in two dimensions, for example)

$$H(z_1, z_2) \;=\; d \cdot H_s \left(\frac{z_1 + z_1^{-1}}{2}, \frac{z_2 + z_2^{-1}}{2} \right),$$

where d denotes a pure delay. The trick is to substitute the one-dimensional kernel $K(z) = (z + z^{-1})/2$ by a multidimensional kernel $K(z_1, z_2)$. As long as this latter kernel is zero-phase, the filter will be linear phase [39]. This method preserves perfect reconstruction and pairs of zeroes ar aliasing frequencies, but works only for linear phase filters.

4. Arbitrary tilings of the time-frequency plane. We have seen in the previous sections the basic facts on one- and multidimensional filter banks and their connection to wavelets. In each case, a specific filter bank/wavelet structure performs a tiling of the time-frequency plane as explained in Section 1. For example, the classical short-time Fourier transform or Gabor transform, and the wavelet transform are just two of many possible tilings. These are illustrated in Figures 4.1(a) and (b) and are similar to what we have discussed in Section 1. We use the term "time-frequency tile" of a particular basis function to designate the region in the plane which contains most of the function's energy. An elegant generalization that contains, at least conceptually, Gabor and wavelet transforms as special cases, is the idea of wavelet packets [7], or, arbitrary subband coding trees. An example of a wavelet packet tiling is given in Figure

4.1(c). While the wavelet packet tiling creates an arbitrary slicing of frequencies (with associated time resolution), it does not change over time. Often a signal is first segmented, and the wavelet packet decomposition is performed on each segment independently. An obvious question is whether we can find a wavelet packet decomposition that changes over time, that is, an arbitrary orthogonal tiling of the time-frequency plane. An example of such a generalized tiling is shown in Figure 4.1(d). We use the term "arbitrary" somewhat casually, since the tiling is restricted to those produced by binary tree structures. However, the wavelet packet construction is generalized sufficiently to warrant the term. In this section, we discuss such arbitrary tilings. By way of illustration, consider the expansion of a

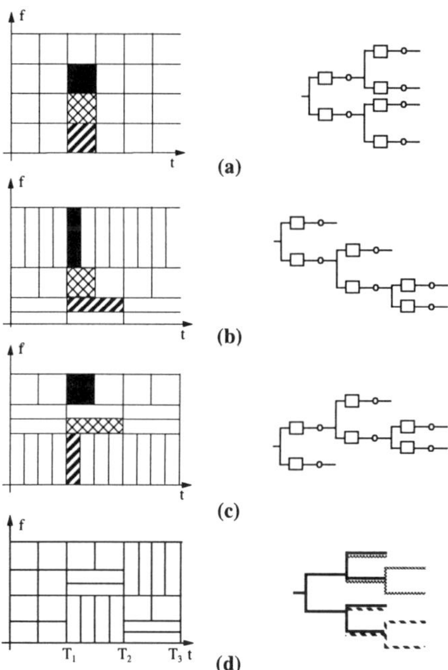

FIG. 4.1. *Tilings of the time-frequency plane. (a) Short-time Fourier transform tiling. (b) Wavelet tiling. (c) Wavelet packet tiling. (d) Generalized tiling which adapts in time as well as in frequency.*

very simple signal using different bases given in Figure 4.2 [18]. The signal consists of 64 points of a sinusoid together with a spike, so it is simple enough to allow depiction of the individual tiles. The shading of a tile in the figures represents the amount of relative energy the coefficient of the corresponding basis function contains. Figure 4.2(a) shows the expansion using the identity basis, where the vectors are the unit sample sequences; here the spike is clearly shown, but the frequency information is not meaningfully revealed. In Figure 4.2(b) the basis is the discrete Fourier transform;

as expected the frequency information of the sinusoid shows very clearly, but the spike does not. A discrete (finite duration) wavelet transform was used in Figure 4.2(c); here both the sinusoid and spike can be seen more or less clearly, and the overall energy is confined to relatively few of the coefficients. If we examine a second signal, however, the picture changes considerably. In Figures 4.3(a), (b) and (c) the same three bases are used to expand another signal, this time consisting of a high-frequency sinusoid and a spike. Figures 4.3(a) and (b) respectively identify the spike and the sinusoid as before, but now the discrete wavelet transform in Figure 4.3 has done poorly also. The reason, of course, is that the wavelet transform has poor frequency localization at high frequencies. Clearly then for many signals this may not be the ideal choice.

The idea of wavelet packets, as we said, is, instead of repeatedly dividing the output of the lowpass channel, to divide at whichever branches of the tree makes sense for the particular signal at hand. This offers more flexibility, since it makes no assumptions about the frequency content of the signal.

The wavelet packet tree structures are still fixed in time. For signals where the frequency characteristics vary with time, it seems obvious that it ought to be possible to do better with a structure that could vary its frequency resolution over time.

Such a decomposition can be realized using the time-varying filter banks which will be presented shortly. They allow one to go between one frequency decomposition and another by merely changing between filter bank trees. This gives us the ability to divide the signal at will in frequency (splitting any frequency band using a two-channel filter bank) and in time (switching between different filter-bank trees) in an orthogonal decomposition. Thus the basis functions that we generate in this way can have energy localizations that give essentially arbitrary tilings of the time-frequency plane.

By way of example Figures 4.2(d) and 4.3(d) show the expansions of the two signals considered before, using time-varying bases. In both cases it is found possible to localize well in both time and frequency provided that the basis can adapt to the signal. The techniques for designing time-varying filter banks that will be presented shortly, allow the freedom to design bases that can, in principle, be adjusted to give parsimonious representations of nonstationary signals. That freedom is only useful, however, if clear rules for deciding when to change the wavelet packet tree can be found. In [37] it was shown how to find the wavelet packet that minimizes the measure J = (Distortion + λ Rate) for a given signal. Since compression is a major application we will use this criterion. The fact that we use two-channel filter banks restricts us to binary splittings of the frequency axis. If we confine ourselves to binary splittings in time also it is possible to find a fast "double-tree" pruning algorithm that finds the best segmentation in time

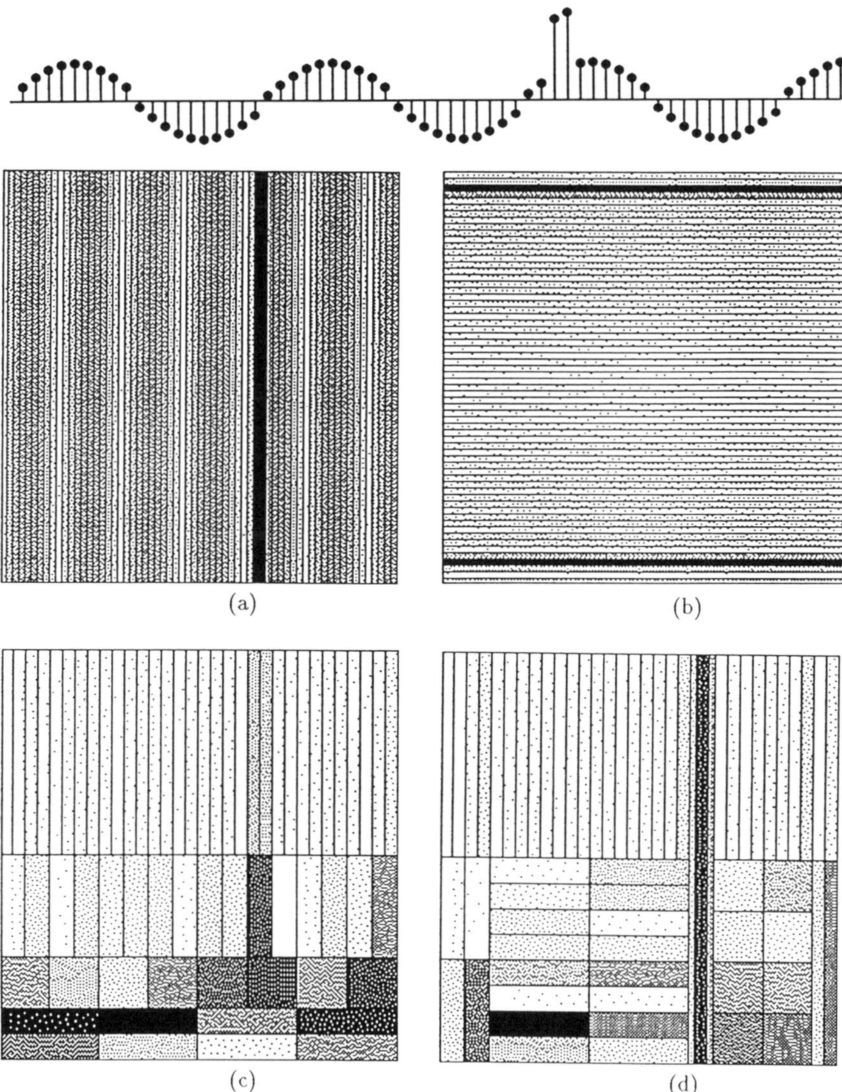

FIG. 4.2. *Tilings for the expansion of a low-frequency sinusoid and a spike using different bases from [18]. (a) Expansion using unit sample sequences as basis functions. (b) Expansion using the discrete Fourier transform. (c) Expansion using the discrete wavelet transform. (d) Expansion based on a time-varying filter bank.*

    C. HERLEY, J. KOVAČEVIĆ, AND M. VETTERLI

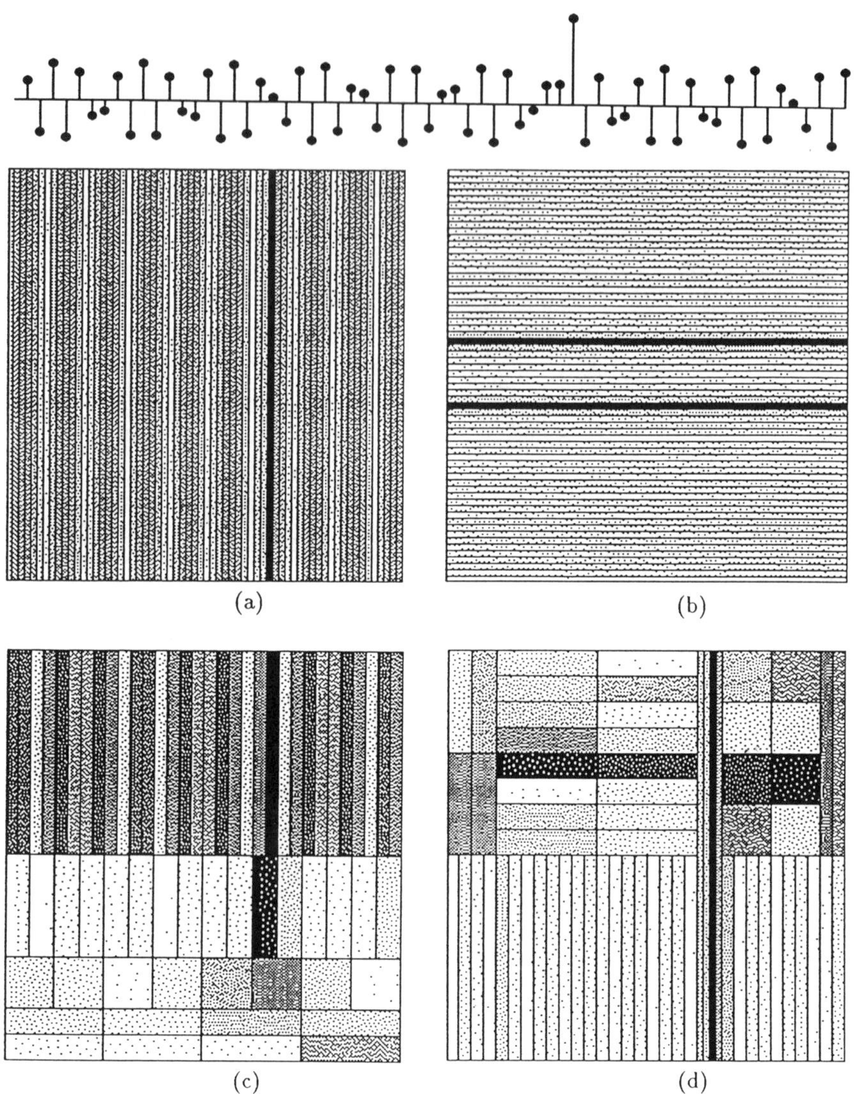

FIG. 4.3. *Tilings for the expansion of a high-frequency sinusoid and a spike using different bases from [18]. (a) Expansion using unit sample sequences as basis functions. (b) Expansion using the discrete Fourier transform. (c) Expansion using the discrete wavelet transform. (d) Expansion based on a time-varying filter bank.*

and frequency simultaneously [19]. In other words, the two optimizations are performed jointly, not sequentially. This gives the time-varying basis that minimizes J, or other additive measure. This holds some promise for compression applications. Note that in this survey we consider only the construction of time-varying filter banks and wavelets, while the algorithm to perform the joint segmentation can be found in [19].

4.1. Time-varying two-channel filter banks. Let us first consider the problem of changing between orthogonal trees based on two-channel filter banks. If we can do this we will be able to construct the most arbitrary tree-based tilings as indicated in Figure 4.1(d). We will make extensive use of the time-domain operator notation for filter banks introduced in Section 2 [34], [48], [50]. Note that the time-varying structures are also studied in [12], [13], [33].

4.1.1. Orthogonalization procedure. The problem of applying an orthogonal filter bank over a finite signal segment involves finding an appropriate way of treating the boundaries. If we take, for example, the case of length-4 filters, applied for the segment $0 \le n \le n_1$, consider the following truncation of the time-domain operator

$$(4.1) \qquad \mathbf{M} = \begin{bmatrix} h_0(1) & h_0(2) & h_0(3) & 0 & 0 & \cdots \\ h_0(2) & -h_0(1) & h_0(0) & 0 & 0 & \cdots \\ 0 & h_0(0) & h_0(1) & h_0(2) & h_0(3) & \cdots \\ 0 & -h_0(3) & h_0(2) & -h_0(1) & h_0(0) & \cdots \\ \vdots & \vdots & \ddots & \ddots & \ddots & \ddots \end{bmatrix}.$$

We have shown the top left corner only, but the bottom right is entirely similar. It is easy to verify that this matrix is square, has full rank, but is no longer unitary. If we denote by M_i the ith row of $\mathbf{M}$ we find that

$$(4.2) \qquad < M_i, M_j > \; = \; 0, \quad i \in \{0,1\}, \quad j \in \{2,3,\cdots n_1 - 1, n_1\};$$

but $< M_0, M_1 > \; \neq 0$ and $< M_{n_1-1}, M_{n_1} > \; \neq 0$. Since the matrix is of full rank, that is, we have a set of linearly independent vectors, we can restore orthogonality using the Gram-Schmidt procedure. To do this start by normalizing the first vector $M_0'' = M_0/\|M_0\|$, and then

$$M_1' \;=\; M_1 - < M_1, M_0'' > M_0'' - \sum_{j=3}^{n_1} < M_1, M_j > M_j$$

$$(4.3) \qquad\qquad =\; M_1 - < M_1, M_0'' > M_0''.$$

The simplification is a consequence of (4.2). Finally set $M_1'' = M_1'/\|M_1'\|$. Note that since M_1 and M_0'' each have only three nonzero entries, so does M_1'' from (4.3). The same procedure is applied to the other boundary vectors M_{n_1-1} and M_{n_1}. A new matrix $\mathbf{M}''$ which has rows

$$\{M_0'', M_1'', M_2, M_3, \cdots, M_{n_1-2}, M_{n_1-1}'', M_{n_1}''\},$$

is then obviously unitary. What is important to note is that $\mathbf{M}''$ has exactly the same zero entries as $\mathbf{M}$; that is, the orthogonal boundary filters have the same support as the truncated filters.

This particularly simple example illustrates a much stronger result, which states that the boundary filters for orthogonal FIR filter banks always have support only in the region of the boundary. We can formally state this as follows:

PROPOSITION 4.1. *The set of boundary filters needed to apply a two-channel orthogonal filter bank, with length-N filters to a finite length signal is a set of $((N-2)/2+d)$ vectors at each boundary, each of which has only $(N-2+d)$ nonzero values. If we define*

$$(4.4) \qquad \mathbf{Q} = \begin{bmatrix} \mathbf{0}_{d_l} & \mathbf{G} & \mathbf{0}_{d_r} \end{bmatrix},$$

where $\mathbf{G}$ is the $(2k \times N_0 + 2(k-1))$ matrix containing the shifted filter impulse responses of the filters $h_0(n)$ and $h_1(n)$, and $\mathbf{0}_{d_l}$ and $\mathbf{0}_{d_r}$ are $(2k \times d_l)$ and $(2k \times d_r)$ matrices of zeros, then the boundary vectors are always of the form

$$\mathbf{e}'_i = (\mathbf{I} - \mathbf{Q}^T \mathbf{Q}) \cdot \mathbf{e}_i,$$

for some $\mathbf{e}_i$.

The proof is given in [20]. The importance of the result is that it is constructive; for any orthogonal filter bank it tells us how to take care of the boundary. This proposition involves carrying out the Gram-Schmidt procedure in operator notation; the novel factor is that the resulting output vectors have nonzero elements only in the region of the transition. The result of the orthogonalization is, of course, not unique. However, given one solution we can explore the space of all possible orthogonal boundary solutions by premultiplying by the matrix

$$(4.5) \qquad \begin{bmatrix} \mathbf{U}_l & \mathbf{0} & \mathbf{0} \\ \mathbf{0} & \mathbf{I}_{n_1-(N-2)-d_l-d_r} & \mathbf{0} \\ \mathbf{0} & \mathbf{0} & \mathbf{U}_r \end{bmatrix},$$

where $\mathbf{U}_l$ and $\mathbf{U}_r$ are unitary matrices of size $((N-2)/2+d_l)$ and $((N-2)/2+d_r)$.

The above orthogonalization procedure is actually a special case. Clearly we can use it to change between orthogonal trees just by calculating an appropriate set of boundary functions on each side of the transition; Proposition 4.1 tells us how to do so. In this case there will be no overlap across the boundary; more general solutions, where there is in fact overlap, are given in [20]. Further, it is possible to use the discrete-time time-varying bases described above in an iterative scheme to derive continuous-time time-varying bases [19], [20], and this is what we discuss next.

4.1.2. Continuous-time bases from discrete-time ones. As we have seen in the previous two sections, we will examine now how the time-varying bases presented above may be used to derive continuous-time ones. This will lead to wavelet bases for interval regions, wavelet bases where the analyzing wavelet varies with time, and wavelet-like bases where the segmentation changes with time. Using a different approach wavelet bases for interval regions have also been investigated by Cohen, Daubechies and Vial [6].

Again, we will rely heavily on matrix notation, and we examine the length-4 case for illustration. Consider the half-infinite block Toeplitz matrix which contains as rows the shifted impulse response of $h_0(n)$, and has the lowpass boundary filter in the first row

$$\mathbf{H}_0 =
\begin{bmatrix}
h_0'(0) & h_0'(1) & h_0'(2) & 0 & 0 & 0 & 0 & 0 & 0 & \cdots \\
0 & h_0(0) & h_0(1) & h_0(2) & h_0(3) & 0 & 0 & 0 & 0 & \cdots \\
0 & 0 & 0 & h_0(0) & h_0(1) & h_0(2) & h_0(3) & 0 & 0 & \cdots \\
\vdots & \vdots & \vdots & \vdots & \vdots & \vdots & \vdots & \vdots & \vdots & \ddots
\end{bmatrix}.$$

(4.6)

Denote by $L_{ik}(z)$ the z-transform of the coefficients of the ith row of $\mathbf{H}_0^k$:

$$L_{ik}(z) = \sum_{n=0}^{\infty} \mathbf{H}_0^k(i, n) z^{-n}.$$

Clearly $L_{11}(z) = H_0'(z)$, $L_{21}(z) = z^{-1} H_0(z)$, $L_{31}(z) = z^{-3} H_0(z)$. To find the $L_{i2}(z)$ we must examine the product $\mathbf{H}_0^2$. We find

$$L_{i2}(z) = z^{-1-4(i-2)} H_0(z) H_0(z^2), \quad i > 1,$$

and, in general

$$(4.7) \qquad L_{ij}(z) = z^{-1-2^j(i-2)} \prod_{k=0}^{j-1} H_0(z^{2^k}) \quad i > 1.$$

The function $L_{2j}(z)$ can easily be recognized as the z-transform of the "graphical iteration" to find the scaling function $\varphi(x)$ of a compactly supported wavelet scheme (see [48]). That is, if we define from $L_{ij}(z)$ a continuous-time function

$$(4.8) \qquad f^{(j)}(x) = l_{2j}(n) \quad n/2^j \le x < (n+1)/2^j,$$

it can be shown that $f^{(j)}(x)$ converges to the scaling function $\varphi(x)$ as $i \to \infty$ (under some constraints on $h_0(n)$). Similarly, the other rows $L_{ij}(z)$, $i > 1$, can be used to converge to $\varphi(x - i - 2)$. For the case of $L_{1j}(z)$ we can define a continuous-time function in a manner similar to (4.8). If this converges as $j \to \infty$ we call it the left-boundary scaling function. The wavelet $\psi(x)$ and the left-boundary wavelet function are found by examining the convergence of the rows of $\mathbf{H}_1 \cdot \mathbf{H}_0^k$. Right boundary functions are found using the right boundary filters. When we deal with Daubechies' filters we use the subscript to denote the order of the filters involved; *i.e.* $\varphi_n(x)$ and $\psi_n(x)$ are the scaling function and wavelet derived from the length-$2n$ Daubechies' filters.

The orthogonality properties that we desire of a wavelet basis (*i.e.* with respect to shift and scale) still hold for systems containing boundary filters. As an illustration observe

$$\mathbf{H}_0^k \cdot (\mathbf{H}_0^*)^k = (\mathbf{H}_0 \cdot \mathbf{H}_0^*)^k = \mathbf{I} \quad \text{and} \quad \mathbf{H}_0^k \cdot (\mathbf{H}_1^*)^k = (\mathbf{H}_0 \cdot \mathbf{H}_1^*)^k = \mathbf{0}.$$

This ensures that the graphical recursions that converge to the wavelet and scaling function are orthogonal with respect to translates at the kth iteration; and this holds true in the limit as $k \to \infty$ also.

Equally, the kth iteration to the wavelet is orthogonal to the $(k-1)$st (*i.e.* its stretched version)

$$(4.9) \qquad \mathbf{H}_1 \cdot \mathbf{H}_0^k \cdot (\mathbf{H}_0^*)^{k-1} \cdot \mathbf{H}_1^* \; = \; \mathbf{H}_1 \cdot \mathbf{H}_0 \cdot \mathbf{H}_1^* \; = \; \mathbf{0}.$$

In the limit this gives orthogonality across scales. More complete and formal statements of these arguments can be found in [20].

4.2. Time-varying modulated lapped transforms. Recall the wavelet packets we mentioned at the beginning of this section (see also Figure 4.1(c)). Obviously, they produce "arbitrary" slicing of frequencies, but do not change over time. Consider what would happen if we would exchange axes in Figure 4.1(c). In this new tiling, the time is sliced up, while the frequency division is uniform, which can be seen as the dual of wavelet packets. Now which system can produce such a tiling? The uniform division of the spectrum can be obtained with many different filter banks structures, one of them being the so-called *modulated lapped transforms (MLT)* [29]. Here we discuss these dual tilings, by using either boundary or overlapping modulated lapped transforms. By boundary modulated lapped transforms, we denote the time-varying tiling of the time-frequency plane where the basis functions do not overlap, while the opposite is true for overlapping modulated lapped transforms, that is, the basis functions from adjacent decompositions do overlap. Although the tilings obtained via time-varying modulated lapped transforms are somewhat restricted as compared to those that can be obtained using the general theory presented in the last section, they offer the advantage that all the filters, both at transitions, and within decompositions, are obtained by modulation. We will call these constructions *time-varying modulated lapped transforms.*

MLT's are a special class of perfect reconstruction filter banks (see, for example, [29]), using a single prototype filter of length $2N$ (where N is the number of channels) to construct all of the filters $h_0, \ldots, h_{N-1}$, by modulation as follows:

$$(4.10) \qquad h_k(n) = \frac{h_{pr}(n)}{\sqrt{N}} \cdot \cos(\frac{(2k+1)}{4N}(2n - N + 1)\pi),$$

with $k = 0, \ldots, N-1$, $n = 0, \ldots, 2N-1$. Here the prototype lowpass filter $h_{pr}(n)$ is usually symmetric and has to satisfy $h_{pr}^2(n) + h_{pr}^2(N-1-n) = 2$. This last condition, imposed on the window, ensures that the resulting MLT is orthogonal. The two symmetric halves of the window are called "tails".

We have seen in Section 2, a convenient way of analyzing filter banks in the time domain, via infinite matrices, such as the one given in (2.12). For modulated lapped transforms, the matrix $\mathbf{A}$ can be written as

$$(4.11) \qquad \mathbf{A} = \begin{bmatrix} \ddots & & & \\ & \mathbf{A}_0 & \mathbf{A}_1 & \\ & & \mathbf{A}_0 & \mathbf{A}_1 \\ & & & \ddots \end{bmatrix},$$

where blocks $\mathbf{A}_0, \mathbf{A}_1$ are of sizes $N \times N$, and contain the impulse responses of the filters. Note that the filter length is twice the number of channels. For example, the jth row of $\mathbf{A}_i$ is $[h_j(2N - 1 - iN) \ldots h_j(N - iN)]$ for $i = 0, 1$. For an orthogonal, perfect reconstruction solution, the matrix $\mathbf{A}$ has to be unitary, which is equivalent to the following [50]:

$$(4.12) \quad \mathbf{A}_0 \mathbf{A}_0^T + \mathbf{A}_1 \mathbf{A}_1^T = \mathbf{I}, \qquad \mathbf{A}_1 \mathbf{A}_0^T = \mathbf{0}, \qquad \mathbf{A}_0 \mathbf{A}_1^T = \mathbf{0}.$$

The second and third conditions in the above are called the "orthogonality of tails" conditions [50].

We want to point out an interesting fact here, that will be used in later constructions. In the matrix $\mathbf{A}$, denote by w_A the window function corresponding to the block $\mathbf{A}_0$. Take the same block, that is, $\mathbf{A}_0$, and change the window function to w_B. Call this newly obtained block $\mathbf{B}_0$. Then $\mathbf{B}_0 \mathbf{B}_0^T = \mathbf{A}_0 \mathbf{A}_0^T$, that is, the product $\mathbf{A}_0 \mathbf{A}_0^T$ does not depend on the window. The same is true of the other block $\mathbf{A}_1$. A consequence of the above, is that the first equation in (4.12) is going to hold, even if we substitute the block $\mathbf{A}_0$ with $\mathbf{B}_0$, or, $\mathbf{A}_1$ with $\mathbf{B}_1$, or, both. In other words, any combination of blocks with different window "tails" will be unitary, or, we will have a nonsymmetric window. Note, however, that this is not sufficient for a valid orthogonal transform, since the "orthogonality of tails" conditions from (4.12) does not hold, namely $\mathbf{A}_0 \mathbf{B}_1^T \neq \mathbf{0}$ (similarly for $\mathbf{B}_0 \mathbf{A}_1^T$). Nevertheless, the fact that $[\mathbf{A}_0 \ \mathbf{B}_1]$ is unitary despite $\mathbf{A}_0$ and $\mathbf{B}_1$ having different windows, is very useful in constructing boundary and overlapping modulated lapped transforms.

As an example, let us construct a set of boundary filters for the modulated lapped transforms. From the results of Section 2, we know that we can always do this by applying the Gram-Schmidt procedure to the appropriately truncated matrix $\mathbf{A}$. However, our approach here is different; namely, we want the boundary filters to be obtained by modulation as well. Therefore, consider the following matrix:

$$(4.13) \qquad \mathbf{A}_b = \begin{bmatrix} \mathbf{B}_0 & \mathbf{A}_1 & & & \\ & \mathbf{A}_0 & \mathbf{A}_1 & & \\ & & \ddots & & \\ & & & \mathbf{A}_0 & \mathbf{A}_1 \\ & & & & \mathbf{A}_0 & \mathbf{B}_1 \end{bmatrix},$$

where $\mathbf{A}_i$ are size-$(N \times N)$ blocks as introduced in (4.11), and $\mathbf{B}_i$ are size-$(N \times N/2)$ blocks with the associated window $w_B(n) = \sqrt{2}$ for $n = N/2, \ldots, 3N/2 - 1$ and zero everywhere else. For example, the ith row of $\mathbf{B}_j$ is given by $[h_{j_B}((3/2)N - 1) \ldots h_{j_B}(N)]$, where $h_{j_B}(n)$ is as given in (4.10) with the window $w_B(n)$.

This demonstrated how to construct boundary filters for a given modulated lapped transform, so as to be able to change decompositions over time. The transform is still orthogonal, by the fact that $\mathbf{A}_b \mathbf{A}_b^T = \mathbf{I}$. There are still N boundary filters, of length $3N/2$. The way they are obtained is by using a nonsymmetric window (which was shown to be possible in the last subsection).

Suppose now that we want to be able to switch between two different modulated lapped transforms, but with filters (basis functions) that overlap. It is possible to change between two same-size modulated lapped transforms but with different windows, as well as change between different-size transforms. Since the basic idea is the same as for boundary filters, we will not discuss it here (for details, see [19]).

REFERENCES

[1] R.ANSARI, *Two-dimensional IIR filters for exact reconstruction in tree-structured subband decomposition*, Electr. Letters **23** (12) June (1987), 633–634.

[2] G.BATTLE, *A block spin construction of ondelettes. Part I: Lemarié functions*, Commun. Math. Phys. **110** (1987), 601–615.

[3] R.N.BRACEWELL, The Fourier Transform and its Applications, McGraw-Hill, New York (second edition) 1965.

[4] A.COHEN, I.DAUBECHIES, *Non-separable bidimensional wavelet bases*, Rev. Mat. Iberoamericana (to appear) (1993).

[5] A.COHEN, I.DAUBECHIES, J.-C.FEAUVEAU, *Biorthogonal bases of compactly supported wavelets*, Commun. on Pure and Appl. Math. **45** (1992), 485–560.

[6] A.COHEN, I.DAUBECHIES, P.VIAL, *Wavelet bases on the interval and fast algorithms*, Journal of Appl. and Comput. Harmonic Analysis **1** (1) Dec (1993), pp. 54–81.

[7] R.Coifman, Y.Meyer, V.Wickerhauser, *Wavelet analysis and signal processing*, Wavelets and their Applications, M.B.Ruska, et al., (eds.) Jones and Bartlett, Boston, MA 1992, 153–178,

[8] R.E.Crochiere, S.A.Weber, J.L.Flanagan, *Digital coding of speech in subbands*, Bell System Technical Journal **55** Oct (1976), 1069–1085.

[9] A.Croisier, D.Esteban, C.Galand, *Perfect channel splitting by use of interpolation/decimation/tree decomposition techniques*, Int. Conf. on Inform. Sciences and Systems, Patras, Greece, August 1976, 443–446.

[10] I.Daubechies, *Orthonormal bases of compactly supported wavelets*, Commun. on Pure and Appl. Math. **41** November (1988), 909–996.

[11] I.Daubechies, Ten Lectures on Wavelets, SIAM, Philadelphia, PA 1992.

[12] R.L.de Queiroz, *Subband processing of finite length signals without border distortions*, Proc. IEEE Int. Conf. Acoust., Speech, and Signal Proc., San Francisco, CA May 1992, 613–616.

[13] R.L.de Queiroz, K.R.Rao, *Time-varying lapped transforms and wavelet packets*, Signal Proc., **Special Issue on Wavelets**, Dec (1993), pp. 3293–3305.

[14] Z.Doïganata, P.P.Vaidyanathan, *Minimal structures for the implementation of digital rational lossles systems*, IEEE Trans. Acoust., Speech, and Signal Proc. **38** (12) Dec (1990), 2058–2074.

[15] A.Gersho, R.M.Gray, Vector Quantization and Signal Compression, Kluwer Academic, Norwell, MA 1992.

[16] I.Gohberg, S.Goldberg, Basic Operator Theory, Birkhauser, Boston, MA 1981.

[17] K.Gröchenig, W.R.Madych, *Multiresolution analysis, Haar bases and self-similar tilings of R^n*, IEEE Trans. Inform. Th., **Special Issue on Wavelet Transforms and Multiresolution Signal Analysis**, March (1992).

[18] C.Herley, Wavelets and Filter Banks (PhD thesis) Columbia University, New York, NY, April 1993.

[19] C.Herley, J.Kovačević, K.Ramchandran, M.Vetterli, *Tilings of the time-frequency plane: construction of arbitrary orthogonal bases and fast tiling algorithms*, IEEE Trans. Signal Proc., **Special Issue on Wavelets and Signal Processing** December (1993), pp. 3341–3359.

[20] C.Herley, M.Vetterli, *Orthogonal time-varying filter banks and wavelet packets*, IEEE Trans. Signal Proc. Oct (1994) (to appear).

[21] C.Herley, M.Vetterli, *Wavelets and recursive filter banks*, IEEE Trans. Signal Proc. Aug (1993), pp. 2536–2556.

[22] J.Kovačević, *Filter banks and wavelets: extensions and applications* (PhD thesis) Columbia University in the City of New York, October 1991.

[23] J.Kovačević, M.Vetterli, *Design of multidimensional non-separable regular filter banks and wavelets*, Proc. IEEE Int. Conf. Acoust., Speech, and Signal Proc., San Francisco, CA **IV** March 1992, 389–392.

[24] J.Kovačević, M.Vetterli, *Non-separable multidimensional perfect reconstruction filter banks and wavelet bases for $\mathcal{R}^n$*, IEEE Trans. Inform. Th., **Special Issue on Wavelet Transforms and Multiresolution Signal Analysis 38** (2) March (1992), 533–555.

[25] W.Lawton, *Application of complex-valued wavelet transforms to subband decomposition*, Signal Proc. Dec (1993), pp. 3566–3568.

[26] W.M.Lawton, H.L.Resnikoff, *Multidimensional wavelet bases*, AWARE (preprint) (1991).

[27] P.G.Lemarié, *Ondelettes à localisation exponentielle*, J. Math. pures et appl. **67** (1988), 227–236.

[28] S.Mallat, *A theory of multiresolution signal decomposition: the wavelet representation*, IEEE Trans. Patt. Recog. and Mach. Intell. **11** (7) July (1989), 674–693.

[29] H.S.Malvar, Signal Processing with Lapped Transforms, Artech House, Norwood, MA 1992.

[30] J.McClellan, *The design of two-dimensional filters by transformations*, Seventh

Ann. Princeton Conf. on ISS, Princeton, NJ 1973, 247–251.

[31] Y.MEYER, Ondelettes, Vol. 1 of Ondelettes et Opérateurs, Hermann, Paris 1990.

[32] F.MINTZER, *Filters for distortion-free two-band multirate filter banks*, IEEE Trans. Acoust., Speech, and Signal Proc. **33** (3) June (1985), 626–630.

[33] K.NAYEBI, T.P.BARNWELL III, M.J.T.SMITH, *Analysis-synthesis systems with time-varying filter bank structures*, Proc. IEEE Int. Conf. Acoust., Speech, and Signal Proc., San Francisco, CA, March 1992, 617–620.

[34] K.NAYEBI, T.P.BARNWELL III, M.J.T.SMITH, *Time domain filter bank analysis: a new design theory*, Signal Proc. **IV** (6) June (1992), 1414–1429.

[35] T.Q.NGUYEN, P.P.VAIDYANATHAN, *Two-channel perfect reconstruction FIR QMF structures which yield linear phase analysis and synthesis filters*, IEEE Trans. Acoust., Speech, and Signal Proc. **37** (5) May 1989, 676–690.

[36] A.PAPOULIS, Signal Analysis, McGraw-Hill, New York, NY 1977.

[37] K.RAMCHANDRAN, M.VETTERLI, *Best wavelet packet bases in a rate-distortion sense*, IEEE Trans. Image Proc. **2** (2) April (1993), 160–175.

[38] T.A.RAMSTAD, *IIR filter bank for subband coding of images*, Proc. IEEE Int. Symp. Circ. and Syst. 1988, 827–830.

[39] I.A.SHAH, A.A.C.KALKER, *Theory and design of multidimensional QMF sub-band filters from 1-D filters using transforms*, IEE Conf. on Image Proc. 1992.

[40] M.J.T.SMITH, *IIR analysis/synthesis systems*, Subband Image Coding, J.W.WOODS (ed.) Kluwer Academic Press, Boston, MA 1991.

[41] M.J.T.SMITH, T.P.BARNWELL III, *A procedure for designing exact reconstruction filter banks for tree structured sub-band coders*, Proc. IEEE Int. Conf. Acoust., Speech, and Signal Proc., San Diego, CA, March 1984.

[42] M.J.T.SMITH, T.P.BARNWELL III, *Exact reconstruction for tree-structured sub-band coders*, IEEE Trans. Acoust., Speech, and Signal Proc. **34** (3) June (1986), 431–441.

[43] P.P.VAIDYANATHAN, Multirate Systems and Filter Banks, Prentice Hall, Englewood Cliffs, NJ 1992.

[44] P.P.VAIDYANATHAN, P.-Q.HOANG, *Lattice structures for optimal design and robust implementation of two-channel perfect reconstruction filter banks*, IEEE Trans. Acoust., Speech, and Signal Proc. **36** (1) January (1988), 81–94.

[45] P.P.VAIDYANATHAN, Z.DOĞANATA, *The role of lossless systems in modern digital signal processing: a tutorial*, IEEE Trans. Educ. **32** (3) August (1989), 181–197.

[46] M.VETTERLI, *Multidimensional subband coding: some theory and algorithms*, Signal Proc. **6** (2) February (1984), 97–112.

[47] M.VETTERLI, *Filter banks allowing perfect reconstruction*, Signal Proc. **10** (3) April (1986), 219–244.

[48] M.VETTERLI, C.HERLEY, *Wavelets and filter banks: theory and design*, IEEE Trans. Signal Proc. **40** (9) September (1992), 2207–2232.

[49] M.VETTERLI, J.KOVAČEVIĆ, *Wavelets and subband coding*, Signal Processing, Prentice-Hall, Englewood Cliffs, NJ (to appear) 1994.

[50] M.VETTERLI, D.J.LEGALL, *Perfect reconstruction FIR filter banks: some properties and factorizations*, IEEE Trans. Acoust., Speech, and Signal Proc. **37** (7) July (1989), 1057–1071.

[51] L.F.VILLEMOES, *Regularity of two-scale difference equations and wavelets*, (PhD thesis) Mathematical Institute, Technical University of Denmark 1992.

[52] J.W.WOODS, S.D.O'NEIL, *Sub-band coding of images*, IEEE Trans. Acoust., Speech, and Signal Proc. **34** (5) October (1986), 1278–1288.

SYSTOLIC ALGORITHMS FOR ADAPTIVE SIGNAL PROCESSING

MARC MOONEN*

Abstract. An overview is given of recent work in parallel algorithms development. It is shown how one specific type of systolic algorithm/array can be used for several 'classical' adaptive signal processing tasks, such as recursive least squares parameter estimation, SVD updating, Kalman filtering, beamforming and direction finding, *etc.*

1. Introduction. Matrix computations play a central role in modern digital signal processing. However, when a real-time implementation is aimed at, the computational complexity involved often represents a serious impediment. Therefore, research has focused on parallel algorithms and architectures, leading to a large collection of different special-purpose architectural constructs, such as, *e.g.*, systolic or wavefront arrays. Very often, different basic algorithmic kernels (*e.g.* basic matrix operations such as matrix-vector products, *etc.*) are implemented separately, and then chained together to obtain the overall solution, see *e.g.* [2]. Here, a different route is taken, by trying to implement different problems on one and the same systolic array. The aim is thus to develop an algorithmic/architectural construct which is useful for a broad class of applications. More specifically, a systolic array is developed which can be used for several 'classical' adaptive signal processing tasks, such as recursive least squares estimation, SVD updating and Kalman filtering. In any case, the efficiency of the resulting systolic implementation is comparable to what is achieved with the best of the existing solutions, if any.

2. Systolic arrays for basic matrix operations. Systolic arrays are processor networks, where in principle each processor performs simple operations on data it receives from its neighbors, and then passes on these data/results to other neighbors. A simple and very popular example is the matrix-vector multiplication array of **Figure 2.1**. Each processor stores one entry of a given matrix V (6×6 in Figure 2.1). A vector a is fed in from the left. The product $\tilde{a}^T = a^T \cdot V$ is accumulated from the bottom to the top, when each processor performs a simple multiply-add operation, as indicated. **Figure 2.2** shows another example, namely the well-known Gentleman-Kung QR updating array [3]. A triangular matrix R is stored and updated with a new data vector (observation). The data vector is fed in from the top, and 'absorbed' in the triangular matrix. This is done by applying a sequence of orthogonal transformations, which are computed on

* ESAT-KUL, K. Mercierlaan 94, 3001 Heverlee, BELGIUM.
e:mail marc.moonen@esat.kuleuven.ac.be. This research was partially sponsored by ESPRIT Basic Research Action Nr. 6632. Marc Moonen is a senior research assistant with the Belgian N.F.W.O. (National Fund for Scientific Research).

the main diagonal and then propagated to the right. For the details, the reader is referred to [3].

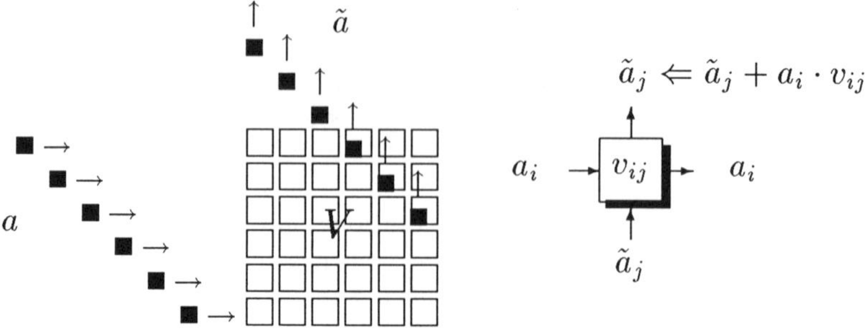

FIG. 2.1. *matrix-vector product*

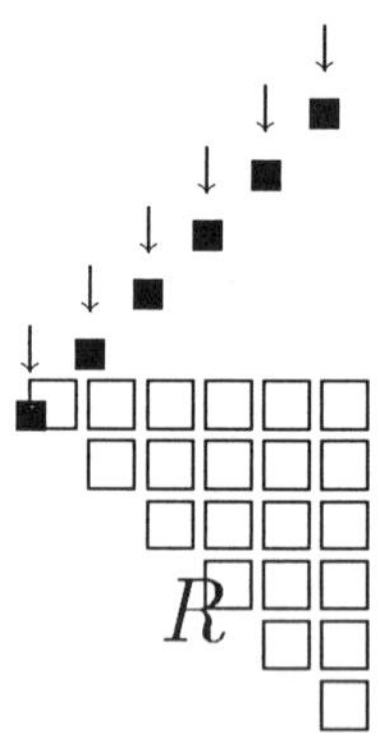

FIG. 2.2. *QR updating array*

An important feature of the above arrays is that successive multiplications/updates can be pipelined. In the Gentleman-Kung array, *e.g.*, as soon as a first update has moved far enough into the array (*i.e.* one step downwards), one can already start with the next update. The throughput is then independent of the size n of R, *i.e.* $\mathcal{O}(n^0)$, which means that in each time step one can start off with a new update. It is important to notice that the array(s) in the next sections also achieve an $\mathcal{O}(n^0)$ throughput rate.

Finally, it should be stressed that so far systolic arrays have not really made their way into many practical applications. Still, the systolic description reveals the fundamental parallelism which is available in an algorithm, and therefore it provides useful information when this algorithm is to be implemented on a present-day parallel architecture. The term 'systolic algorithm' is often used here, to emphasize the algorithmic nature rather than the specific architectural style.

3. A combined array for SVD updating. Our aim is now to move on to more complicated problems. Many practical signal processing algorithms consist of a combination of several basic algorithmic kernels, *e.g.* basic matrix operations such as matrix-vector products, matrix decompositions, *etc.* In this section, the SVD updating example of [6,7] is used to introduce an array which has turned out to be useful for a whole class of applications. In the next section, a few other applications are shown.

<u>SVD and SVD updating</u>

The singular value decomposition (SVD) of a given (real) matrix A is given as

$$\underbrace{A}_{m \times n} = \underbrace{U}_{m \times n} \cdot \underbrace{\Sigma}_{n \times n} \cdot \underbrace{V}_{n \times n}.$$

where $U^T U = I$, $V^T V = I$ and Σ is a diagonal matrix. It is used in systems identification, harmonic retrieval, beamforming, *etc.* Here we consider the SVD updating problem, where the aim is to track the matrices Σ and V, when new observations are added in each time step, *i.e.* when A is defined in a recursive manner

$$A_k = \left[\frac{A_{k-1}}{a_k^T} \right].$$

An algorithm for this is developed and analyzed in [6]. The idea is to combine QR updating with a so-called Jacobi-type SVD algorithm [5]. An orthogonal matrix V is stored and updated, together with a triangular matrix R, which is always close to Σ [1].

In the algorithmic description below, the triangular matrix R (close to a diagonal matrix), is indicated by $\begin{bmatrix} x & \cdots \\ & x & \vdots \\ & & x \end{bmatrix}$, for the sake of clarity. Initially, $R = 0$ and $V = I$. The algorithm is then given as follows :

$\qquad$ <u>for</u> $k = 1, \ldots, \infty$

---

[1] When R is close to Σ, the V-matrix is also close to the true V-matrix. In [6] it is shown that the tracking error is always smaller than the time variation in $\mathcal{O}(n)$ time steps. The tracking error and the time variation are defined in terms of the angles between true and estimated subspaces or subspaces at different time steps.

1. matrix-vector multiplication

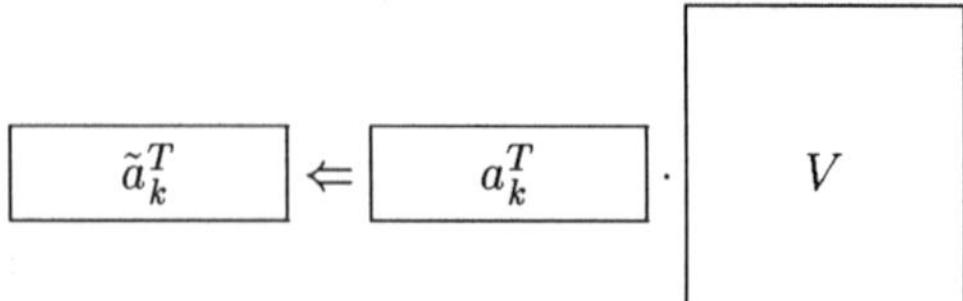

2. QR update

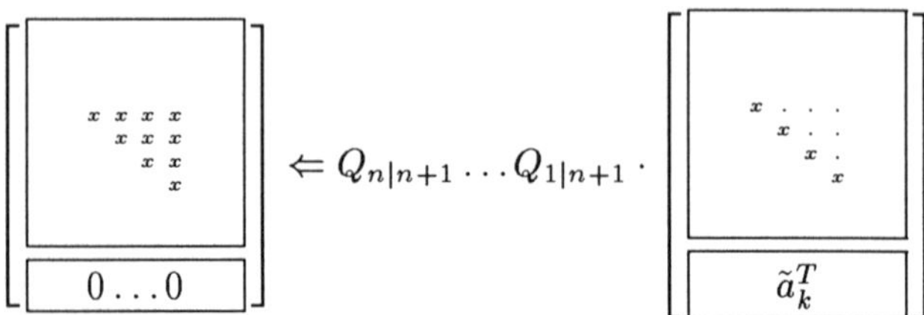

3. SVD steps (Jacobi)

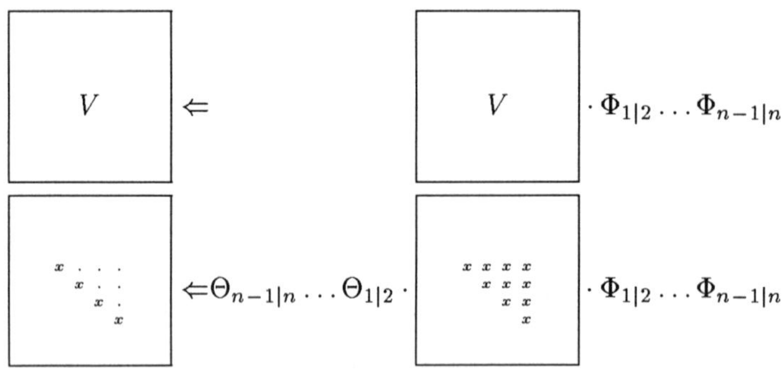

<u>end</u>

The QR update in step 2 is performed with a sequence of n plane transformations (row transformations). $Q_{1|n+1}$ combines rows 1 and $n+1$ of the right-hand matrix, to zero the $(n+1,1)$ entry. $Q_{2|n+1}$ then combines rows 2 and $n+1$ to zero the $(n+1,2)$ entry, *etc.* The QR update moves the R matrix further from the diagonal form. The diagonal form is then (partly) restored in step 3, where a sequence of row and column transformations is applied (Jacobi-type diagonalization). $\Theta_{1|2}$ and $\Phi_{1|2}$ combine rows 1 and 2 and columns 1 and 2 respectively, so that the (1,2) entry is zeroed, *etc.* One sequence of $n-1$ row/column transformations is applied after each QR update. For more details, we refer to [5,6].

<u>SVD updating array</u>

The above algorithm combines matrix-vector products with QR updates and the Jacobi-type diagonalization process. The aim is now to come up with a systolic array. The 'backbone' is the SVD array of [5], given in **Figure 3.1**, which will be combined with the arrays of Figure 2.1 and 2.2. In Figure 3.1, R and V are stored on top of each other (dots represent matrix entries). In Figure a, all 2×2 blocks on the main diagonal with so-called 'odd pivot indices' are picked, and corresponding row and column (plane) transformations are computed. Column transformations are then propagated upwards, row transformations are propagated to the right, Figure b. As soon as this first transformation 'front' has moved far enough, the next one is started off, Figure c. Now all 2×2 blocks with 'even pivot indices' are picked. After Figure d comes Figure a again. For more details, we refer to [5,6].

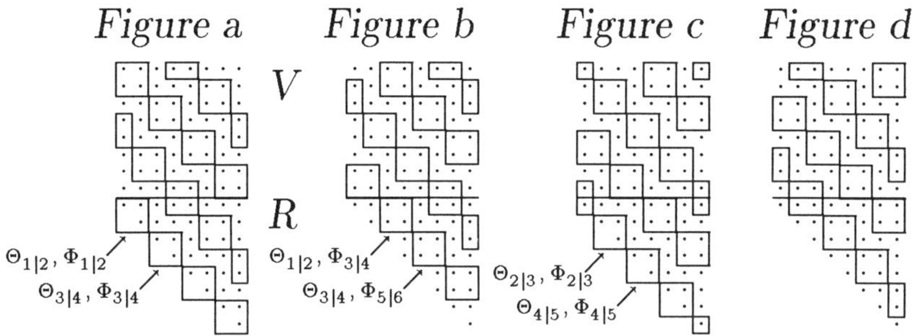

Fig. 3.1. *SVD array*

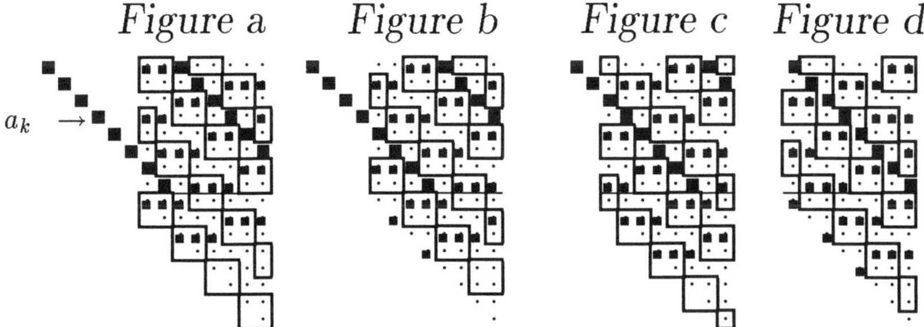

Fig. 3.2. *SVD updating array*

The array of Figure 3.1 is then combined with the arrays of Figures 2.1 and 2.2. The resulting array is depicted in **Figure 3.2**. The new data vectors are fed in from the left. The matrix vector products are computed in the upp er square part (indicated with the ■ $- s$). The results are available at the top end of the array, and then propagated downwards, towards the triangular array (indicated with the ■ $- s$). The QR updating is performed in the triangular array. The Jacobi-type diagonalization process is operated simultaneously. A new data vector is fed in after each 'odd-even cycle', *i.e.* the throughput is $\mathcal{O}(n^0)$. For the details, we refer to [7].

What is remarkable about this array, is that combining the different computational steps (matrix-vector products, QR updating and SVD diagonalization) is really done at the 'processor level' (2×2 level). This is clear from the functional description of the 2×2 blocks in **Figure 3.3** and **Figure 3.4**. Each processor performs a 2×2 matrie-vector product (only in the square part), a 2×2 QR related operation (only in the triangular part), and a 2×2 SVD related operation. In other words, the arrays of Figures 2.1, 2.2 and 3.1 are truly overlaid/combined instead of simply being chained.

<u>INTERNAL NODE V-MATRIX</u>

1. matrix-vector product (if pivot index = odd)

$$\blacksquare \leftarrow \blacksquare + \oplus \cdot a_p + \ominus \cdot a_{p+1}$$
$$\bar{\blacksquare} \leftarrow \bar{\blacksquare} + \otimes \cdot a_p + \oslash \cdot a_{p+1}$$

2. Apply transformation (SVD step)

$$\begin{bmatrix} \bullet & \bar{\bullet} \\ \oplus & \otimes \\ \ominus & \oslash \end{bmatrix} \quad \leftarrow \quad \begin{bmatrix} \bullet & \bar{\bullet} \\ \oplus & \otimes \\ \ominus & \oslash \end{bmatrix} \begin{bmatrix} \cos\phi & \sin\phi \\ -\sin\phi & \cos\phi \end{bmatrix}$$

3. Propagate $\blacksquare$, $\bar{\blacksquare}$

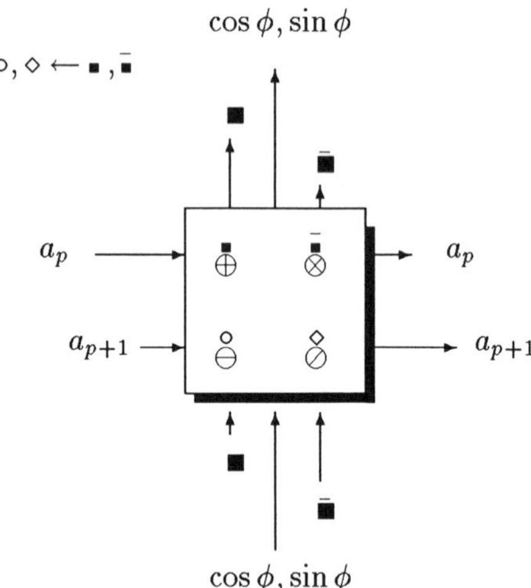

FIG. 3.3. *Functional description for an internal node in the V matrix*

INTERNAL NODE R-MATRIX

1. Apply transformation (SVD step)

$$\begin{bmatrix} \blacksquare & \bar{\blacksquare} \\ \oplus & \otimes \\ \ominus & \oslash \end{bmatrix} \leftarrow \begin{bmatrix} 1 & \\ & \cos\theta \ \sin\theta \\ & -\sin\theta \ \cos\theta \end{bmatrix} \begin{bmatrix} \blacksquare & \bar{\blacksquare} \\ \oplus & \otimes \\ \ominus & \oslash \end{bmatrix} \begin{bmatrix} \cos\phi \ \sin\phi \\ -\sin\phi \ \cos\phi \end{bmatrix}$$

2. Apply transformation (QR update)

$$\begin{bmatrix} \blacksquare & \bar{\blacksquare} \\ \oplus & \otimes \\ \ominus & \oslash \end{bmatrix} \leftarrow \begin{bmatrix} \cos\psi \ \sin\psi & \\ -\sin\psi \ \cos\psi & \\ & 1 \end{bmatrix} \begin{bmatrix} \blacksquare & \bar{\blacksquare} \\ \oplus & \otimes \\ \ominus & \oslash \end{bmatrix}$$

3. Propagate $\blacksquare, \bar{\blacksquare}$

$$\circ, \diamond \leftarrow \blacksquare, \bar{\blacksquare}$$

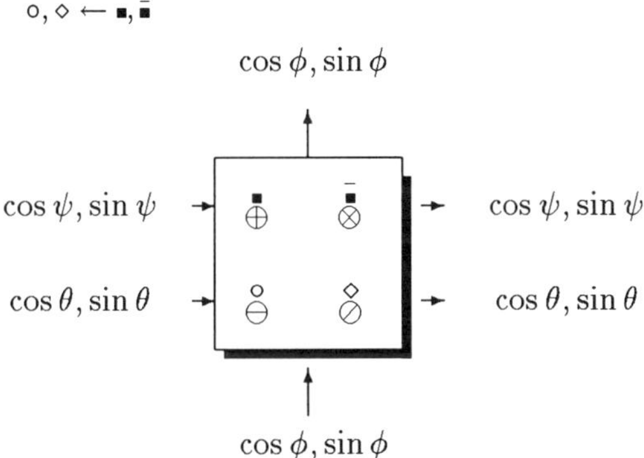

FIG. 3.4. *Functional description for an internal node in the R matrix*

4. Other applications. While SVD updating is important in its own right, the systolic algorithm/array of the previous section turns out to be useful for many other adaptive signal processing tasks, too. Here, only two examples are given, namely recursive least squares estimation and Kalman filtering. For other applications we refer to [8,9,13].

Recursive least squares estimation

The least squares problem is given as

$$\min_x \| \underbrace{A}_{m \times n} \cdot x - \underbrace{y}_{m \times 1} \|_2$$

132   MARC MOONEN

where A and y are given (real) matrices. With

$$Q^T\,[\,A|y\,] = \left[\begin{array}{c|c} R & z \\ \hline 0 & \zeta \end{array}\right]$$

where $Q^T Q = I$, the solution is given as

$$x_{LS} = R^{-1}z.$$

Here, the recursive least squares (RLS) problem is considered, where A and y are defined in a recursive manner

$$A_k = \left[\begin{array}{c} A_{k-1} \\ a_k^T \end{array}\right] \qquad y_k = \left[\begin{array}{c} y_{k-1} \\ \gamma_k \end{array}\right].$$

The 'standard' RLS algorithm is based on QR updates and triangular back-solves, but it is well known that these two steps cannot be pipelined on a systolic array. In [9], an RLS algorithm is given that works with the triangular factor R *and* its inverse. The systolic description corresponds to the array of the previous section, but it is quite involved. Another RLS algorithm, taken from [15] and which works with R^{-1} only, is turned into a systolic algorithm in [10,11]. The systolic description is slightly different and slightly more involved, compared to the SVD array. Here, an alternative RLS algorithm is given which directly fits onto the SVD array. A square matrix $P^{-1} = \square$ is stored and updated, such that $P_k^T P_k = A_k^T A_k$, together with the least squares solution x_{LS} :

    <u>for</u> $k = 1,\ldots,\infty$

1. matrix-vector multiplication

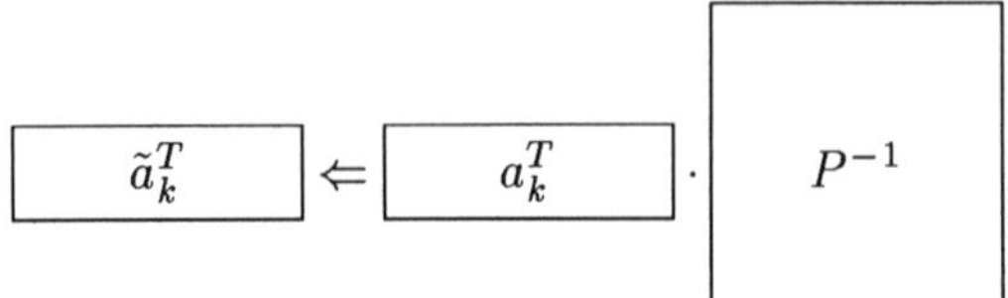

2. orthogonal update

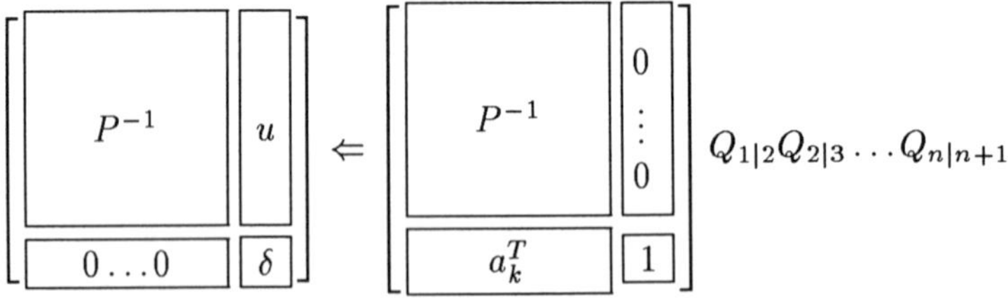

3. update LS solution

$$x_{LS} \Leftarrow x_{LS} - (\gamma_k - a_k^T x_{LS})\cdot \frac{u}{\delta}$$

<u>end</u>

The above algorithm is similar to the algorithm of [15]. The only difference is that it works with a square matrix, instead of a triangular matrix. In particular, this allows to reorder the transformations in step 2, which leads to a simpler systolic implementation. In step 2, $Q_{1|2}$ combines columns 1 and 2, such that the $(n+1, 1)$ element in the compound matrix is zeroed. $Q_{2|3}$ combines columns 2 and 3, such that the $(n+1, 2)$ element is zeroed, *etc.* **Figure 4.1** shows how this is implemented on the SVD array. The square part stores and updates the factor P^{-1}. The triangular part does not store any matrix, but it is used to compute the orthogonal transformations. The matrix vector products (step 1) are computed in the square part. The results are propagated from the top to the triangular part, where $Q_{1|2}$, $Q_{2|3}$, *etc.* are computed (step 2). These transformations are then propagated upwards. One additional column should be added to the right, to store and update x_{LS} and to perform the last transformation $Q_{n|n+1}$. This is not shown in Figure 4.1 (but trivial). The throughput is again $\mathcal{O}(n^0)$, which means that data vectors are fed in and least squares solutions are computed at a rate which is independent of n. The functionality of the 2×2 blocks is slightly different from Figure 3.2, but easily derived.

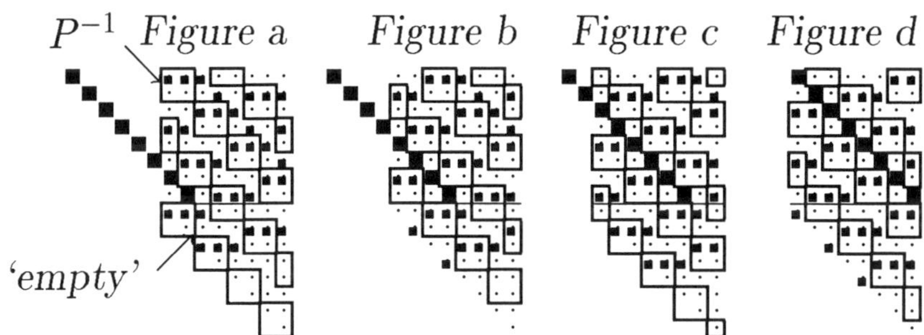

FIG. 4.1. *RLS array*

<u>Kalman filtering</u>

The Kalman filter is a fundamental tool for state estimation in such areas as signal processing and modern control. Suppose we are given a linear multivariable and time-varying system, with a state space model as follows

$$x_{k+1} = A_k x_k + B_k u_k + w_k$$
$$y_k = C_k x_k + D_k u_k + v_k.$$

The aim is to estimate the state x_k at each time k. The matrices A_k,

B_k, C_k and D_k are the known system matrices at time k. Vectors u_k, y_k and x_k denote the deterministic forcing vector (m-vector), the measurement vector (l-vector) and the state vector at time k. The dimension of x_k is the system order n. Only an initial expected value for x_0 is available, namely $x_{0|-1}$, together with its error covariance matrix $E\{(x_0 - x_{0|-1})(x_0 - x_{0|-1})^T\} = P_{0|-1}^{\frac{1}{2}} P_{0|-1}^{\frac{T}{2}}$. The 'square root' $P_{0|-1}^{\frac{1}{2}}$—or more generally $P_{k|k-1}^{\frac{1}{2}}$—is upper triangular here. Finally, w_k and v_k are unknown noise sequences—process noise and measurement noise—with zero mean and known covariances $E\{v_k v_k^T\} = V_k^{\frac{1}{2}} V_k^{\frac{T}{2}}$, $E\{w_k w_k^T\} = W_k^{\frac{1}{2}} W_k^{\frac{T}{2}}$. The aim of the Kalman filter is to provide an estimate for the state vector, by making use of observations of the in- and outputs. The 'predicted state estimate' $x_{k+1|k}$, which is used here, is the estimate of the state vector x_{k+1}, given observations up until time k.

Following the work by Duncan and Horn [1] and Paige and Saunders [14], the square root information filter (predictor) is defined as follows, and employs a single orthogonal transformation

$$
Q_k \cdot \left[\begin{array}{cc|c}
P_{k|k-1}^{-\frac{1}{2}} & 0 & P_{k|k-1}^{-\frac{1}{2}} x_{k|k-1} \\
\hline
W_k^{-\frac{1}{2}} A_k & -W_k^{-\frac{1}{2}} & -W_k^{-\frac{1}{2}} B_k u_k \\
\hline
V_k^{-\frac{1}{2}} C_k & 0 & V_k^{-\frac{1}{2}}(y_k - D_k u_k)
\end{array}\right]
$$
$$
\underbrace{}_{\text{prearray}}
$$

$$
= \left[\begin{array}{cc|c}
* & * & * \\
\hline
0 & P_{k+1|k}^{-\frac{1}{2}} & P_{k+1|k}^{-\frac{1}{2}} x_{k+1|k} \\
\hline
0 & 0 & *
\end{array}\right] \cdot
$$
$$
\underbrace{}_{\text{postarray}}
$$

The first block row in the prearray, which allows to compute $x_{k|k-1}$, is propagated from the previous time step. The second and third block rows correspond to the state space equations for time step k. The orthogonal transformation Q_k triangularizes this prearray (QR factorization). Apart from a number of 'don't care' entries, one then has an updated square root $P_{k+1|k}^{-\frac{1}{2}}$ together with a corresponding right-hand side $P_{k+1|k}^{-\frac{1}{2}} x_{k+1|k}$. The state estimate $x_{k+1|k}$ may then be computed by backsubstitution. Both $P_{k+1|k}^{-\frac{1}{2}}$ and $P_{k+1|k}^{-\frac{1}{2}} x_{k+1|k}$ are finally propagated to the next time step, $etc.$

In [12], it is shown that this can be recast as follows (some of the details are omitted) :

$$
R \Leftarrow \left[\begin{array}{cc|c}
0 & 0 & 0 \\
\hline
0 & P_{k|k-1}^{-\frac{1}{2}} & P_{k|k-1}^{-\frac{1}{2}} x_{k|k-1}
\end{array}\right]
$$

$$\Pi \Leftarrow I$$

<u>for</u> $j = 1, \ldots, n + l$

1. matrix-vector product

$$a_j^T \Leftarrow j\text{-th row of } \left[\begin{array}{c|c|c} 0 & V_k^{-\frac{1}{2}} C_k & V_k^{-\frac{1}{2}} (y_k - D_k u_k) \\ \hline -W_k^{-\frac{1}{2}} & W_k^{-\frac{1}{2}} A_k & -W_k^{-\frac{1}{2}} B_k u_k \end{array}\right]$$

$$\boxed{\tilde{a}_j^T} \Leftarrow \boxed{a_j^T} \cdot \boxed{\Pi}$$

2. QR update

$$\left[\frac{R}{0}\right] \Leftarrow Q \cdot \left[\frac{R}{\tilde{a}_j^T}\right]$$

3. column permutations/row transformations iff $j > l$:

$$\Pi \Leftarrow \Pi \cdot \Pi_{1/2} \Pi_{2/3} \ldots \Pi_{2n-1/2n}$$
$$R \Leftarrow \Theta_{2n-1/2n} \ldots \Theta_{2/3} \Theta_{1/2} \cdot R \cdot \Pi_{1/2} \Pi_{2/3} \ldots \Pi_{2n-1/2n}$$

<u>end</u>

$$\left[\begin{array}{c|c|c} * & * & * \\ \hline 0 & P_{k+1|k}^{-\frac{1}{2}} & P_{k+1|k}^{-\frac{1}{2}} x_{k+1|k} \end{array}\right] \Leftarrow R$$

Here Π is a permutation matrix. The state estimate can be computed by adding a few simple computations, see [12]. The above algorithm is seen to consist of the same computational steps as the SVD updating algorithm, which is remarkable (only the transformations in step 3 are computed in a different way). As a result, the SVD array can be used to implement a Kalman filter. This is outlined in **Figure 4.2**. The state equations are fed in from the left, the state estimates run out at the bottom. The operations for subsequent time steps are easily pipelined. This Jacobi-type Kalman filter is roughly as efficient as *e.g.* Kung's array [4], which is the most efficient array known in the literature. For the details, we refer to [12]

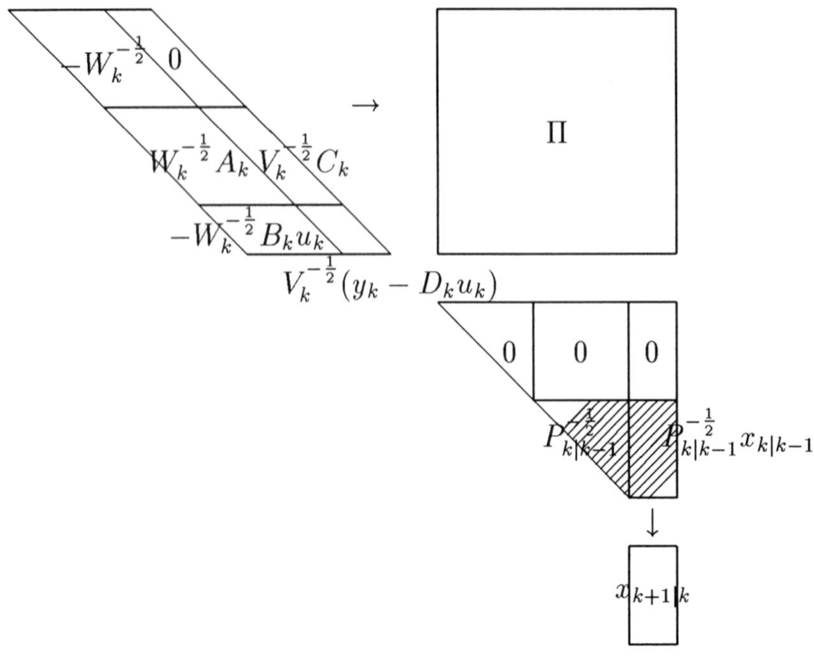

FIG. 4.2. *Kalman filter*

5. Conclusions. It is shown how one specific type of systolic algorithm/array can be used for several 'classical' adaptive signal processing tasks, such as recursive least squares parameter estimation, SVD updating and Kalman filtering. This is important in view of eventual hardware implementation. It suffices to tune one of the above algorithms to the type of parallel architecture one has available, to be able to implement all the other algorithms.

REFERENCES

[1] D.B.DUNCAN, S.D.HORN, *Linear dynamic recursive estimation from the viewpoint of regression analysis*, J. Amer. Statist. Assoc. **67** (1972), 815–821.

[2] F.GASTON, G.IRWIN, *Systolic kalman filtering: an overview*, IEE Proceedings **137** (4) (1990), 235–244.

[3] W.M.GENTLEMAN, H.T.KUNG, *Matrix triangularization by systolic arrays*. Real-Time Signal Processing IV, Proc. SPIE **298** (1982), 19–26.

[4] S.Y.KUNG, *VLSI array processors*, Englewood Cliffs, NJ., Prentice Hall 1988.

[5] F.T.LUK, *A triangular processor array for computing singular values*. Lin. Alg. Appl. **77** (1986), 259–273.

[6] M.MOONEN, P.VAN DOOREN, J.VANDEWALLE, *An SVD updating algorithm for subspace tracking*, Internal Report K.U. Leuven, ESAT/SISTA 1989-13. (to appear in) SIAM J. Matrix Anal. Appl. **13** (4) (1992).

[7] M.MOONEN, P.VAN DOOREN, J.VANDEWALLE, *A systolic array for SVD updating*, Internal Report K.U. Leuven, ESAT/SISTA 1990-18. (to appear in) SIAM J. Matrix Anal. Appl. (1993).

[8] M.MOONEN, P.VAN DOOREN, J.VANDEWALLE, *A systolic algorithm for QSVD updating*. Signal Processing **25** (2) (1991), 203–213.

[9] M.MOONEN, J.VANDEWALLE, *Recursive least squares with stabilized inverse factorization*, Signal Processing **21** (1) (1990), 1–15.

[10] M.MOONEN, J.VANDEWALLE, *A systolic array for recursive least squares computations*, Internal Report K.U. Leuven, ESAT/SISTA 1990-22. (to appear in) IEEE Trans. Signal Processing, 1993.

[11] M.MOONEN, J.VANDEWALLE, *A square root covariance algorithm for constrained recursive least squares estimation*, Journal of VLSI Signal Processing **3** (3) (1991), 163–172.

[12] M. MOONEN, *Implementing the square-root information Kalman filter on a Jacobi-type systolic array*. Internal Report K.U. Leuven, ESAT/SISTA 1991-30. (to appear in) Journal of VLSI Signal Processing.

[13] M.MOONEN, F.VAN POUCKE, E.DEPRETTERE, *Parallel and adaptive high resolution direction finding*. Internal Report K.U. Leuven, ESAT/SISTA 1992-32. (submitted for publication).

[14] C.C.PAIGE, M.SAUNDERS, *Least squares estimation of discrete linear dynamic systems using orthogonal transformations*, SIAM J. Numer. Anal. **14** (2) (1977), 180–193.

[15] C.T.PAN, R.J.PLEMMONS, *Least squares modifications with inverse factorization: parallel implications*, Journal of Computational and Applied Mathematics **27** (1-2) (1989), 109–127.

ADAPTIVE ALGORITHMS FOR BLIND CHANNEL EQUALIZATION

JOHN G. PROAKIS*

Abstract. Several different approaches to the design of blind channel equalization algorithms for digital communications have been described in the literature, including steepest-descent algorithms, algorithms based on the use of high-order statistics, and algorithms based on the maximum-likelihood criterion. In this paper, we focus on algorithms based on maximum likelihood optimization for jointly estimating the channel characteristics and the data sequence.

1. Introduction. In high speed data communication systems, intersymbol interference (ISI) caused by channel amplitude and phase distortion requires channel equalization in order to make a correct decision as to which data symbol is transmitted. Conventionally, equalization is done first through a training mode, where a known data sequence is transmitted for initial adjustment of the equalizer parameters, and is then followed by a decision-directed scheme for tracking any time variations in the channel characteristics. However, problems arise in multipoint networks and multipath fading channels, where the receiver has to perform equalization of the channel without a training mode. When the receiver is "blind" to a training data sequence, the problem is known as blind equalization.

A number of adaptive algorithms have been proposed for blind equalization [1–18]. Most of these algorithms are based on the use of steepest descent algorithms for adaptation of the equalizer [1–8]. Others are based on the use of higher-order statistics of the received signal to estimate the channel characteristics and to design the equalizer [9–12]. More recently, blind equalization algorithms based on the maximum likelihood criterion have been proposed [13–18].

In this paper, we consider the problem of joint channel estimation and data detection based on the maximum likelihood criterion. In the following section, we formulate the blind equalization problem based on maximum likelihood optimization. Then we describe algorithms for performing the optimization.

2. Formulation of maximum likelihood optimization. Figure 2.1 illustrates the discrete-time model of the digital communication system under consideration. The input to the channel is a sequence $\{a_n\}$ of signal points taken from either a PAM or a QAM signal constellation. The channel is modeled as a linear FIR filter having $L + 1$ symbol-spaced taps, where L represents the span of the ISI and $\{h_k\}$ denote the channel impulse response coefficients. The output of the channel is corrupted by a

* CDSP Research Center, Department of Electrical and Computer Engineering, Northeastern University, Boston, Massachusetts 02115. This work was supported in part by the National Science Foundation under grant MIP-9115526.

noise sequence $\{w_n\}$, which is characterized as a white zero mean, Gaussian sequence having a variance σ^2.

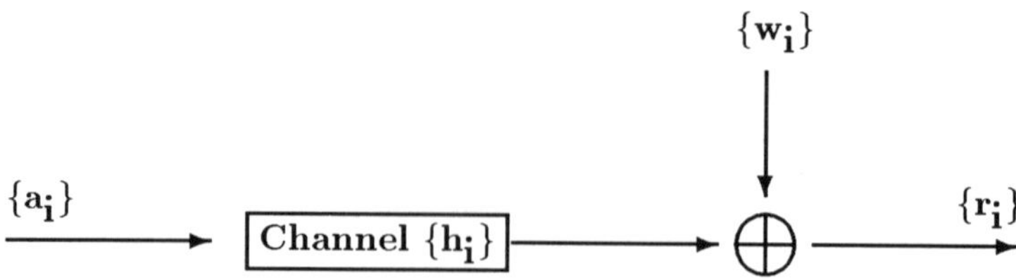

FIG. 2.1. *Discrete model of a communication system*

Hence, the received sequence may be expressed as

$$(2.1) \qquad r_n = \sum_{k=0}^{L} h_k a_{n-k} + w_n, \quad n = 1, 2 \ldots$$

For a block of N received data points, the (joint) probability density function of the received data vector $\mathbf{r} = (r_1, r_2, \ldots, r_N)^t$ conditioned on knowing the impulse response vector $\mathbf{h} = (h_0, h_1, \ldots, h_L)^t$ and the data vector $\mathbf{a} = (a_1, a_2, \ldots, a_N)^t$ is

$$(2.2) \qquad p(\mathbf{r}|\mathbf{h}, \mathbf{a}) = \frac{1}{(2\pi\sigma^2)^N} e^{-\frac{1}{2\sigma^2} \sum_{n=1}^{N} |r_n - \sum_{k=0}^{L} h_k a_{n-k}|^2}$$

where the superscript t denotes the transpose of a vector or matrix. The joint maximum likelihood estimate of $\mathbf{h}$ and $\mathbf{a}$ are the values of these vectors that maximize the joint probability density function $p(\mathbf{r}|\mathbf{h}, \mathbf{a})$ or, equivalently, the values of $\mathbf{h}$ and $\mathbf{a}$ that minimize the term in the exponent. Hence, the ML solution is simply the minimum over $\mathbf{h}$ and $\mathbf{a}$ of the cost function

$$(2.3) \qquad \begin{aligned} J(\mathbf{a}, \mathbf{h}) &= \sum_{n=1}^{N} \left| r_n - \sum_{k=0}^{L} h_k\, a_{n-k} \right|^2 \\ &= \left\| \mathbf{r} - \mathbf{A}\, \mathbf{h} \right\|^2 \end{aligned}$$

where the matrix $\mathbf{A}$ is called the **data matrix** and is defined as

$$(2.4) \qquad \mathbf{A} = \begin{pmatrix} a_1 & 0 & 0 & \ldots & 0 \\ a_2 & a_1 & 0 & \ldots & 0 \\ a_3 & a_2 & a_1 & \ldots & 0 \\ \vdots & \vdots & \vdots & & \vdots \\ a_N & a_{N-1} & a_{N-2} & \ldots & a_{N-L} \end{pmatrix}$$

We make several observations. First of all, we note that when the data vector $\mathbf{a}$ (or the data matrix $\mathbf{A}$) is known, as is the case when a training sequence is available at the receiver, the ML channel impulse response estimate obtained by minimizing 2.3 over $\mathbf{h}$ is

$$(2.5) \qquad \mathbf{h}_{ML}(\mathbf{a}) = \left(\mathbf{A}^t \mathbf{A}\right)^{-1} \mathbf{A}^t\, \mathbf{r}$$

On the other hand, when the channel impulse response $\mathbf{h}$ is known, the optimum ML detector for the data sequence $\mathbf{a}$ performs a trellis search (or tree search) by utilizing the well known Viterbi algorithm for the ISI channel.

When neither $\mathbf{a}$ nor $\mathbf{h}$ are known, the minimization of the performances index $J(\mathbf{a},\ \mathbf{h})$ may be performed jointly over $\mathbf{a}$ and $\mathbf{h}$. Alternatively, $\mathbf{h}$ may be estimated from the probability density function $p(\mathbf{r}|\mathbf{h})$, which may be obtained by averaging $p(\mathbf{r}, \mathbf{a}|\mathbf{h})$ over all possible data sequences. That is,

$$(2.6) \qquad \begin{aligned} p\left(\mathbf{r}|\mathbf{h}\right) &= \sum_m p\left(\mathbf{r}, \mathbf{a}^{(m)}|\mathbf{h}\right) \\ &= \sum_m p\left(\mathbf{r}|\mathbf{a}^{(m)}, \mathbf{h}\right)\, P\left(\mathbf{a}^{(m)}\right) \end{aligned}$$

where $P(\mathbf{a}^{(m)})$ is the probability of the sequence $\mathbf{a} = a^{(m)}$, for $m = 1, 2, \ldots, M^N$ and M is the size of the signal constellation.

Below, we describe several algorithms based on these ML optimization methods.

3. An algorithm based on an average over data sequences. As indicated in the above discussion, when both $\mathbf{a}$ and $\mathbf{h}$ are unknown, one approach is to estimate the impulse response $\mathbf{h}$ after averaging the probability density $p(\mathbf{r}, \mathbf{a}|\mathbf{h})$ over all possible data sequences. Thus, we have

$$(3.1) \qquad \begin{aligned} p\left(\mathbf{r}|\mathbf{h}\right) &= \sum_m p\left(\mathbf{r}|\mathbf{a}^{(m)}, \mathbf{h}\right) P\left(\mathbf{a}^{(m)}\right) \\ &= \sum_m \left[\tfrac{1}{(2\pi\sigma^2)^N} e^{-\|\mathbf{r}-\mathbf{A}^{(m)}\mathbf{h}\|^2/2\sigma^2}\right] P\left(\mathbf{a}^{(m)}\right) \end{aligned}$$

Then, the estimate of $\mathbf{h}$ that maximizes $p(\mathbf{r}|\mathbf{h})$ is the solution of the equation

$$(3.2) \qquad \begin{aligned} \tfrac{\partial p(\mathbf{r}|\mathbf{h})}{\partial \mathbf{h}} = \sum_m P\left(\mathbf{a}^{(m)}\right) \\ \left(\mathbf{A}^{(m)t}\mathbf{A}^{(m)}\mathbf{h} - \mathbf{A}^{(m)t}\mathbf{r}\right) e^{-\|\mathbf{r}-\mathbf{A}^{(m)}\mathbf{h}|^2} = 0 \end{aligned}$$

Hence, the estimate of $\mathbf{h}$ may be expressed as

$$(3.3) \qquad \begin{aligned} \mathbf{h} = \Bigg[&\sum_m P\left(\mathbf{a}^{(m)}\right) \mathbf{A}^{(m)t}\mathbf{A}^{(m)} g\left(\mathbf{r}, \mathbf{A}^{(m)}, \mathbf{h}\right) \Bigg]^{-1} \\ &\sum_m P\left(\mathbf{a}^{(m)}\right) g\left(\mathbf{r}, \mathbf{A}^{(m)}, \mathbf{h}\right) \mathbf{A}^{(m)t}\mathbf{r} \end{aligned}$$

where the function $g(\mathbf{r}, \mathbf{A}^{(m)}, \mathbf{h})$ is defined as

$$(3.4) \qquad g\left(\mathbf{r}, \mathbf{A}^{(m)}, \mathbf{h}\right) = e^{-\|\mathbf{r} - \mathbf{A}^{(m)}\mathbf{h}\|^2 / 2\sigma^2}$$

The resulting solution for the optimum $\mathbf{h}$ is denoted as $\mathbf{h}_{ML}$.

Equation 3.3 is a nonlinear equation for the estimate of the channel impulse response, given the received signal vector $\mathbf{r}$. It is generally difficult to obtain the optimum solution by directly solving 3.3. On the other hand, it is relatively simple to devise a numerical method that solves for $\mathbf{h}_{ML}$ recursively. Specifically, we may write

$$\mathbf{h}^{(k+1)} = \left[\sum_m P(\mathbf{a}^{(m)})\mathbf{A}^{(m)t}\mathbf{A}^{(m)} g(\mathbf{r}, \mathbf{A}^{(m)}, \mathbf{h}^{(k)})\right]^{-1}$$
$$(3.5) \qquad \sum_m P\left(\mathbf{a}^{(m)}\right) g\left(\mathbf{r}, \mathbf{A}^{(m)}, \mathbf{h}^{(k)}\right) \mathbf{A}^t \mathbf{r}$$

Once $\mathbf{h}_{ML}$ is obtained from the solution of 3.3 or 3.5, we may simply use the estimate in the minimization of the metric $J(\mathbf{a}, \mathbf{h}_{ML})$, given by 2.3, over all the possible data sequences. Thus, $\mathbf{a}_{ML}$ is the sequence $\mathbf{a}$ that minimizes $J(\mathbf{a}, \mathbf{h}_{ML})$, i.e.,

$$(3.6) \qquad \min_{\mathbf{a}} J\left(\mathbf{a}, \mathbf{h}_{ML}\right) = \min_{\mathbf{a}} \|\mathbf{r} - \mathbf{A}\mathbf{h}_{ML}\|^2$$

We know that the Viterbi algorithm is the computationally efficient algorithm for performing the minimization of $J(\mathbf{a}, \mathbf{h}_{ML})$ over $\mathbf{a}$.

This algorithm has two major drawbacks. First, the recursion for $\mathbf{h}_{LM}$ given by 3.5 is computationally intensive. Secondly, and, perhaps, more importantly, the estimate $\mathbf{h}_{ML}$ is not as good as the maximum likelihood estimate $\mathbf{h}_{ML}(\mathbf{a})$ that is obtained when the sequence $\mathbf{a}$ is known. Consequently, the error rate performance of the blind equalizer (the Viterbi algorithm) based on the estimate $\mathbf{h}_{ML}$ is poorer than that based on $\mathbf{h}_{ML}(\mathbf{a})$.

In the following sections we describe maximum-likelihood algorithms based on joint optimization over $\mathbf{a}$ and $\mathbf{h}$.

4. Generalized viterbi algorithm for joint channel and data estimation. In this section, we consider the joint optimization of the performance index $J(\mathbf{a}, \mathbf{h})$ given by 2.3. Since the elements of the impulse response vector $\mathbf{h}$ are continuous and the elements of the data vector are discrete, one approach is to determine the maximum likelihood estimate of $\mathbf{h}$ for each possible data sequence and, then, to select the data sequence that minimizes $J(\mathbf{a}, \mathbf{h})$ for each corresponding channel estimate. Thus, the channel estimate corresponding to the mth data sequence $\mathbf{a}^{(m)}$ is

$$(4.1) \qquad \mathbf{h}_{ML}\left(\mathbf{a}^{(m)}\right) = \left(\mathbf{A}^{(m)t}\mathbf{A}^{(m)}\right)^{-1} \mathbf{A}^{(m)t}\mathbf{r}.$$

For the mth data sequence, the cost function $J(\mathbf{a}, \mathbf{h})$ becomes

$$(4.2) \qquad J\left(\mathbf{a}^{(m)}, \mathbf{h}_{ML}\left(\mathbf{a}^{(m)}\right)\right) = \| \mathbf{r} - \mathbf{A}^{(m)}\mathbf{h}_{ML}\left(\mathbf{a}^{(m)}\right) \|^2$$

Then, from the set of M^N possible sequences, we select the data sequence that minimizes the cost function in 4.2, i.e., we determine

$$(4.3) \qquad \min_{\mathbf{a}^{(m)}} \; J\left(\mathbf{a}^{(m)}, \mathbf{h}_{ML}\left(\mathbf{a}^{(m)}\right)\right)$$

The approach described above is an exhaustive computational search method with a computational complexity that grows exponentially with the length of the data block. Since $N \geq L$, this method becomes impractical when the length of the ISI span L and the block length N are large. Consequently, it is desirable to find computationally efficient algorithms that constrain the search for finding the minimum over $\mathbf{a}^{(m)}$ and, simultaneously, reduce the number of channel estimates that are obtained.

Seshadri [15] devised a computationally efficient algorithm for performing the joint optimization over a reduced set of data sequences. In essence, Seshadri's algorithm is a type of generalized Viterbi algorithm (GVA) that retains $K \geq 1$ best estimates of the transmitted data sequence into each state of the trellis and the corresponding channel estimates. To elaborate, we first recall that the conventional Viterbi algorithm (VA) retains the most probable data sequence at each state of the trellis. The remaining sequences are discarded. Hence, for an M-ary signal constellation and a channel length $L + 1$, the number of states is M^L and the number of surviving sequences is also M^L. This is the case when the channel impulse response is known and the metric computations are based on the known channel characteristic. On the other hand, when the channel impulse response $\mathbf{h}$ is unknown, it becomes necessary to retain more than one candidate sequence in each state.

In Seshadri's GVA, the search is identical to the conventional VA from the beginning up to the L - stage of the trellis, i.e, up to the point where the received sequence $(r_1, r_2, \ldots, r_L)$ has been processed. Hence, up to the L - stage, an exhaustive search is performed. Associated with each data sequence $\mathbf{a}^{(m)}$, there is a corresponding channel estimate $\mathbf{h}_{ML}(\mathbf{a}^{(m)})$. From this stage on, the search is modified to retain $K \geq 1$ surviving sequences and associated channel estimates per state instead of only one sequence per state. Thus, the GVA is used for processing the received signal sequence $\{r_n, \; n \geq L + 1\}$. The channel estimate is updated recursively at each stage using the LMS algorithm to further reduce the computational complexity. Simulation results given in the paper by Seshadri [15] indicate that this GVA blind equalization algorithm performs rather well at moderate signal-to-noise ratios with $K = 4$. Hence, there is a modest increase in the computational complexity of the GVA compared with that for the conventional VA. However, there are additional computations involved with the estimation and updating of the channel estimates $\mathbf{h}(\mathbf{a}^{(m)})$ associated with each of the surviving data estimates.

Below, we describe an alternative joint optimization algorithm that avoids the least-squares computation for channel estimation. Instead, we work with a quantized channel model of length $L + 1$ and use the conventional VA to determine the maximum likelihood data sequence.

5. Quantized-channel algorithm. We may reverse the order in which we perform the joint minimization of the performance index $J(\mathbf{a}, \mathbf{h})$ given by 2.3. That is, we may select a channel impulse response, say $\mathbf{h} = \mathbf{h}^{(1)}$ and then use the conventional VA to find the optimum sequence for this channel impulse response. Then, we may modify $\mathbf{h}^{(1)}$ in some manner to $\mathbf{h}^{(2)} = \mathbf{h}^{(1)} + \Delta\mathbf{h}^{(1)}$ and repeat the optimization over the data sequences $\{\mathbf{a}^{(m)}\}$.

Based on this general approach, we have developed a new ML blind equalization algorithm, which we call a **quantized - channel algorithm**. The algorithm operates over a grid in the channel space, which becomes finer and finer by using the ML criterion to confine the estimated channel in the neighborhood of the original unknown channel. This algorithm leads to an efficient parallel implementation and its storage requirements are only those of the VA.

Algorithm description. Throughout this section we assume that we know the order of the channel $L + 1$ and its energy. That is

$$(5.1) \qquad \mathcal{E}_h = \sum_{l=0}^{L} |h_l|^2$$

Although the second assumption is realistic, in most practical cases, the true order of the channel is generally unknown. Below, we show how one can cope with the unknown order of the channel. In addition, we assume that binary antipodal signaling is employed. The generalization to M-ary signal constellations is straightforward.

For real channels of order $L + 1$, there are 2^{L+1} different sign combinations, one of which matches the signs of the taps of the channel. Thus, as an initial set of candidate channels we assume the set

$$(5.2) \qquad \mathcal{H}_1 = \left\{ \beta\mathbf{b}_j, \qquad j = 0, \ldots, 2^{L+1} - 1 \right\}$$

where $\mathbf{b}_j$ is the vector corresponding to the binary expansion of the index j, with -1 replacing 0. The parameter β is adjusted in such a way that the energy of the candidate channels equals the energy of the unknown channel. That is

$$(5.3) \qquad \beta^2(L + 1) = \mathcal{E}_h$$

The Viterbi algorithm is then used, to find the channel that produces the smallest accumulated error energy. It is conceivable at this point, that if a channel $\beta\mathbf{b}_i$ is retained due to its low cost, then the channel $-\beta\mathbf{b}_i$ is also a candidate for further processing, due to the symmetry of the signal constellation. This ambiguity is immaterial if the input bits are differentially encoded, so the number of the initial candidate channels can be reduced by a factor of two.

Suppose now that we have decided in favor of the channel $\mathbf{h}^{(1)} = \beta \mathbf{b}_i$ with associated cost $\mathcal{E}^{(1)}$, that is

$$(5.4) \qquad \mathcal{E}^{(1)} = \min_{\mathbf{a}} ||\mathbf{r} - \beta \mathbf{b}_i * \mathbf{a}||^2$$

The cost is compared with the initial $\mathcal{E}^{(0)} = ||\mathbf{r}||^2$, and if it is greater than this, we repeat the process after halving β. If $\mathcal{E}^{(1)}$ is less than $\mathcal{E}^{(0)}$, then we consider a new set of candidate channels as follows

$$(5.5) \qquad \mathcal{H}_2 = \left\{ \mathbf{h}^{(1)} + \beta \mathbf{b}_j, \qquad j = 0, \ldots, 2^{L+1} - 1 \right\}$$

Again we find $\mathbf{h}^{(2)}$ with the smallest $\mathcal{E}^{(2)}$, and we compare the latter with $\mathcal{E}^{(1)}$. If $\mathcal{E}^{(2)} > \mathcal{E}^{(1)}$ we halve β and we repeat the process. If $\mathcal{E}^{(2)} < \mathcal{E}^{(1)}$ we continue with the set

$$(5.6) \qquad \mathcal{H}_3 = \left\{ \mathbf{h}^{(2)} + \beta \mathbf{b}_j, \qquad j = 0, \ldots, 2^{L+1} - 1 \right\}$$

The process is terminated after a fixed number (C) of halving operations of the step β. **Figure 5.1** shows a flow chart for the proposed algorithm. In every step, the algorithm tries to maximize the likelihood function by selecting the channel with the smallest accumulated error energy.

The major advantage of the algorithm is its parallel structure, since the search of the 2^{L+1} candidate channels can be performed simultaneously over the same block of data. When the signal-to-noise ratio (SNR) is low, a good strategy is to run each family of candidate channels over different blocks of data. In this way, if the algorithm is trapped in the neighborhood of a false channel (because of the noise) it is very unlikely that it will remain in the vicinity of this channel if we switch to another block of data.

The major disadvantage of the algorithm is its complexity, which increases exponentially with the order of the channel. The number of iterations is not that crucial as we will see in the next section. After the second or the third halving operation of the step β, the recovered data sequence $\mathbf{a}$ is highly correlated with the true sequence and thus it can be used directly to estimate the channel, using for example a LS algorithm.

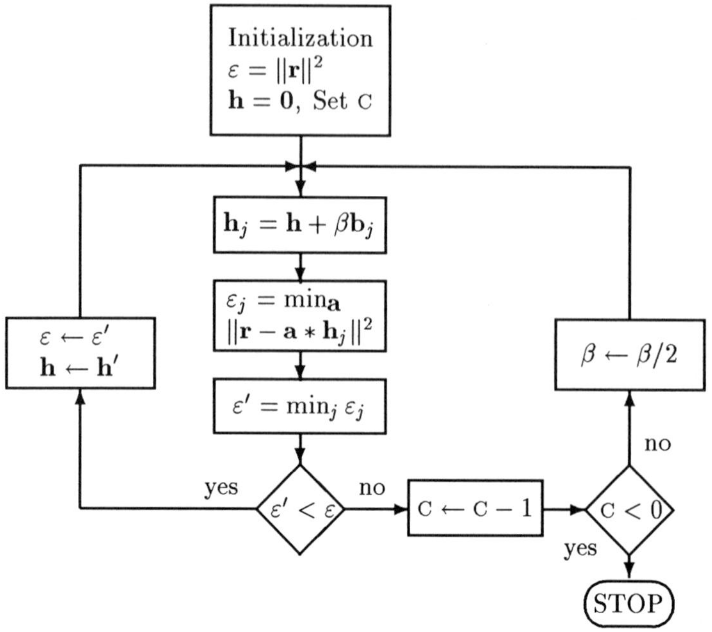

FIG. 5.1. *Flow-chart of the algorithm*

6. Simulation results. We have tested the proposed algorithm with three channels taken from Proakis [19]. The impulse response of the channels is given by

$$
\begin{aligned}
\mathbf{h}_a &= \quad [.407 \quad .815 \quad .407] \\
\mathbf{h}_b &= \quad [-.21 \quad .-.50 \quad .72 \quad .36 \quad .21] \\
\mathbf{h}_c &= \quad [.227 \quad .460 \quad .688 \quad .460 \quad .227]
\end{aligned}
\tag{6.1}
$$

The initial step β was 0.6 for the first channel and 0.45 for the others. The length N of the data blocks was 50, and the algorithm was run over the same data block after halving the step β.

Table 6.1 shows the coefficients of channel (a) and their estimates just before each halving of the step β. The constellation was PAM-2 and the results were obtained after averaging over 100 independent data blocks. The SNR, measured with respect to the output of the channel, was set to 30 and 10 db. The last two columns of the table refer to the mean square error

$$
J_{\mathbf{h}} = E\left\{ \|\mathbf{h}_a - \mathbf{h}\|^2 \right\}
$$

for the cases of 30 db and 10 db SNR, respectively.

Table 6.2 and Table 6.3 show the results for channels (b) and (c). For channel (b) 4 out of 100 runs, in the case of 30 db SNR, and 23 runs in the case of 10 db SNR converged to wrong channels. For these cases we had to increase the length of the data blocks from 50 to 100 samples. The latter suggests that channels with severe phase distortion are more difficult to identify with short data blocks.

When the alphabet size is A, the number of the states of the trellises is A^L, and this may be prohibitive for large A. A reduced constellation approach can be used with multilevel signaling, as was proposed by Sato [5], [6].

Table 6.4 refers to channels (b), (c), with an input constellation PAM-8 and SNR equal to 30 db. The length of the data blocks was set to 200 for channel (c), and to 400 for channel (b). The algorithm operated with trellises of $S = 2^{L+1}$ states, assuming that the transmitted signal was binary with values ± 4. As it is observed, identification of the channels is still possible within hundreds of symbols.

So far, we have assumed that we know the channel memory L. When the order of the channel is unknown, the method in [15] can be used for the order mismatch. Table 6.5 refers to a typical example of overparameterizing the channel. Channel (a) is used with hypothesized order $L = 5$, and an input constellation PAM-2. The initial step β equals .45 and SNR = 30 db. The estimated channel is shown for 5 independent blocks of length 50 samples. The results were obtained after 5 halvings of the step β. As it is observed the channel is identifiable, subject to sign inversions, delays or advances which are immaterial to the blind equalization problem. When we underparameterize the channel, the algorithm captures the dominant coefficients of the channel. In this case we can pad zeros to the left and right side of the estimated impulse response and re-run the algorithm with a smaller β. Table 6.6 refers to channels (b) and (c) with an hypothesized order $L = 3$, and initial β equal to .6. The length of the data blocks was set to 100 for channel (c) and to 300 for channel (b). The previous results indicate a possibility to reduce the computational complexity. One can start with a low-order channel and a large step β. After the dominant coefficients have been estimated we can extend the channel and use a fine grid (small β) for the remaining coefficients.

β	$h_{a,0}$	$h_{a,1}$	$h_{a,2}$	$J_{\mathbf{h}}$ 30 db	$J_{\mathbf{h}}$ 10 db
	.407	.815	.407		
..6	.600	.600	.600	.1207	.1402
..3	.300	.900	.300	.0301	.0552
..15	.450	.750	.450	.0079	.0368
..075	.375	.825	.375	.0021	.0347
..0375	.411	.790	.411	.0008	.0300

TABLE 6.1
Channel (a), PAM-2, N=50, SNR=30 db, 10 db

β	$h_{b,0}$	$h_{b,1}$	$h_{b,2}$	$h_{b,3}$	$h_{b,4}$	$J_{\mathbf{h}}$ 30 db	$J_{\mathbf{h}}$ 10 db
	-.21	-.50	.72	.36	.21		
..450	-.450	-.450	.450	.450	.450	.1987	.2025
..225	-.225	-.670	.675	.225	.225	.0517	.0685
..1125	-.225	-.434	.675	.441	.225	.0140	.0280
..0562	-.186	-.503	.714	.377	.188	.0052	.0190
..0281	-.210	-.492	.717	.358	.210	.0012	.0176

TABLE 6.2
Channel (b), PAM-2, N=50, SNR=30 db, 10 db

β	$h_{c,0}$	$h_{c,1}$	$h_{c,2}$	$h_{c,3}$	$h_{c,4}$	$J_{\mathbf{h}}$ 30 db	$J_{\mathbf{h}}$ 10 db
	.227	.460	.688	.460	.227		
..450	.450	.450	.450	.450	.450	.1563	.1974
..225	.225	.513	.648	.504	.225	.1121	.1830
..1125	.222	.436	.650	.438	.222	.0207	.1064
..0562	.222	.434	.650	.436	.222	.0198	.1018
..0281	.222	.436	.655	.438	.222	.0187	.1039

TABLE 6.3
Channel (c), PAM-2, N=50, SNR=30 db, 10 db

β	Channel (b) $J_{\mathbf{h}}$	Channel (c) $J_{\mathbf{h}}$
..6	.1987	.1563
..3	.0888	.1003
..15	.0552	.0638
..975	.0440	.0680
..0375	.0450	.0640

TABLE 6.4
Channel (b), (c), PAM-8, SNR=30 db

Block No.	$h_{c,0}$	$h_{c,1}$	$h_{c,2}$	$h_{c,3}$	$h_{c,4}$
1	.393	.843	.393	.000	.000
2	.000	.393	.843	.393	.000
3	.000	.393	.843	.393	.000
4	.000	.000	-.393	-.787	-.393
5	.000	.393	.843	.393	.000
6	.393	.843	.393	.000	.000
7	.393	.843	.393	.000	.000
8	.393	.843	.393	.000	.000
9	.393	.843	.393	.000	.000
10	.393	.787	.393	.000	.000

TABLE 6.5
Channel (a), PAM-2, N=50, SNR=30 db, L=5

Block No.	Channel (b)			Channel (c)		
1	-.487	.712	.337	.450	.675	.450
2	-.487	.712	.337	.450	.675	.450
3	-.487	.712	.337	.450	.675	.450
4	-.487	.712	.337	.450	.675	.450
5	-.487	.712	.337	.487	.712	.487
6	-.487	.712	.337	.450	.675	.450
7	-.487	.712	.337	.450	.675	.450
8	-.487	.712	.337	.225	.450	.450
9	-.487	.712	.337	.487	.712	.487
10	-.487	.712	.337	.450	.675	.450

TABLE 6.6
Channel (b), (c), PAM-2, SNR=30 db, L=3

7. Conclusion. We have described blind equalization algorithms based on the maximum likelihood criterion for jointly estimating the channel impulse response and the data sequence. Compared to other methods, this approach to blind equalization has the advantage of being optimal and requires relatively few received signal samples for performing channel estimation. However, the computational complexity of the algorithms is large when the intersymbol interference (ISI) spans many symbols. On some channels, such as the mobile radio channel, where the span of the ISI is relatively short, these algorithms are simple to implement.

REFERENCES

[1] S.BELLINI, *Bussgang techniques for blind equalization*, Proceedings of IEEE-Globecom'86 1986, 46.1.1–46.1.7.

[2] A.BENVENISTE, M.GOURSAT, *Blind equalizers*, IEEE Trans. on Communications **COM-32** August (1984), 871–883.

[3] D.N.GODARD, *Self-recovering equalization and carrier tracking in two-dimensional data communication systems*, IEEE Trans. on Communications **COM-28** November (1980), 1867–1875.

[4] G.PICCHI, G.PRATI, *Blind equalization and carrier recovery using a "stop-and-go" decision-directed algorithm*, IEEE Trans. on Communications **COM-35** (9) September (1987), 877–887.

[5] Y.SATO, *A method for self-recovering equalization for multilevel amplitude-modulation systems*, IEEE Trans. on Communications, **COM-23** June (1975), 679–682.

[6] Y.SATO, ET AL, *Blind suppression of time dependency and its extension to multi-dimensional equalization*, Proc. of IEEE ICC'86 1986, 46.4.1–46.4.5.

[7] O.SHALVI, E.WEINSTEIN, *New criteria for blind equalization of nonminimum phase systems (channels)*, IEEE Trans. on Information Theory, March 1990.

[8] J.R.TREICHLER, B.G.AGEE, *A new approach to multipath correction of constant modulus signals*, IEEE Trans. on Acoustics, Speech, and Signal Processing **ASSP-31** (2) April (1983), 459–471.

[9] D.HATZINAKOS, C.L.NIKIAS, *Blind equalization using a tricepstrum based algorithm*, IEEE Trans. on Communications, May 1991.

[10] A.G.BESSIOS, C.L.NIKIAS, *POTEA: The power cepstrum and tricoherence equalization algorithm*, IEEE Transactions on Signal Processing (under review) 1991.

[11] B.PORAT B.FRIEDLANDER, *Blind equalization of digital communication channels using higher-order moments*, IEEE Trans. on Signal Processing **39** February (1991), 522–526.

[12] R.PAN, C.L.NIKIAS, *The complex cepstrum of higher-order cumulants and non-minimum phase system identification*, IEEE Trans. ASSP **ASSP-36** February (1988), 186–205.

[13] G.KAWAS, R.VALLET, *Joint detection and estimation for transmission over unknown channels*, Proc. Douzienne Colloque GRETSI, Juan-Les-Pins, France, June 12-16, 1989.

[14] 14 M.FEDER, J.A.CATIPOVIC, *Algorithms for joint channel estimation and data recovery—application to equalization in underwater communications*, Journal of Ocean Engineering, January (1991).

[15] N.SESHADRI, *Joint data and channel estimation using blind trellis search techniques* (submitted for publication) 1991.

[16] M.GHOSH, C.L.WEBER, *Maximum-likelihood blind equalization*, Proc. 1991 SPIE Conference, San Diego, CA, July 22-26, 1991.

[17] E.Zervas, J.G.Proakis, V.Eyuboglu, *Effects of constellation shaping on blind equalization*, Proc. SPIE Conference, San Diego, CA, July 22-26, 1991.

[18] E.Zervas, J.G.Proakis, V.Eyuboglu, *A quantized channel approach to blind equalization*, Proc. ICC'91, Chicago, ILL, June 15-17, 1991.

[19] J.G.Proakis, *Digital communications*, McGraw-Hill 1989.

SQUARE-ROOT ALGORITHMS FOR STRUCTURED MATRICES, INTERPOLATION, AND COMPLETION PROBLEMS

A.H. SAYED[*], T. CONSTANTINESCU[†], AND T. KAILATH[‡]

Abstract. We derive square-root based algorithms for structured matrices and discuss potential applications to interpolation and matrix completion problems. The mathematical machinery used here is based on a standard Gaussian elimination technique and on simple results from matrix and linear system theory. We show how to exploit the inherent displacement structure in order to construct a convenient transmission-line cascade that makes evident the required interpolation and completion conditions. We also introduce the concept of time-variant structured matrices and discuss its applications to matrix completions and time-variant interpolation problems.

1. Introduction. Interpolation problems of various types have had many applications in circuit and system theory. A classical paper is that of Youla and Saito [1], which was followed up and significantly extended by Helton [2] and others. We refer to the works of Sarason [3], Adamjan, Arov, and Krein [4], Foias and Frazho [5], Fedčina [6], Delsarte, Genin, and Kamp [7], Ball and Helton [8], Alpay, Dewilde, and Dym [9,10,11], Kimura [12], Ball, Gohberg, and Rodman [13], Limebeer, Anderson, and Green [14,15], and others, for extensive discussion and references.

The successful application of interpolation problems in control and circuit theory has inspired the study of generalizations to the time-variant setting [16,17,18,19,20,21]. We describe here a computationally oriented solution for interpolation problems, in both the time-variant and time-invariant cases, based on a fast algorithm for the recursive triangular factorization of structured matrices. We use the interpolation data to construct a convenient so-called generator for the factorization algorithm, which then leads to a transmission-line cascade of first-order sections that makes evident the interpolation property. This is due to the fact that transmission lines have "transmission zeros": certain inputs at certain frequencies yield zero outputs. In the time-invariant case for example, each section of the cascade can be characterized by a $(p + q) \times (p + q)$ rational transfer matrix $\Theta_i(z)$

---

[*] Dept. of Electrical and Computer Engineering, University of California, Santa Barbara, CA 93106. email: sayed@ece.ucsb.edu. fax: (805)893-3262. phone: (805)893-4457. This work was supported in part by a fellowship from Fundação de Amparo à Pesquisa do Estado de São Paulo and by Escola Politécnica da Universidade de São Paulo, BRAZIL. This work was also supported in part by the Air Force Office of Scientific Research, Air Force Systems Command under Contract AFOSR91-0060, and by the Army Research Office under contract DAAL03-89-K-0109.

[†] Programs in Mathematical Sciences, University of Texas at Dallas, Richardson, TX 75083.

[‡] Information Systems Laboratory, Stanford University, Stanford, CA 94305.

say, that has a left zero-direction vector g_i at a frequency f_i, viz.,

$$g_i\Theta_i(f_i) \equiv \begin{bmatrix} a_i & b_i \end{bmatrix} \begin{bmatrix} \Theta_{i,11} & \Theta_{i,12} \\ \Theta_{i,21} & \Theta_{i,22} \end{bmatrix}(f_i) = \mathbf{0},$$

which makes evident (with the proper partitioning of the row vector g_i and the matrix function $\Theta_i(z)$) the following interpolation property: $a_i\Theta_{i,12}\Theta_{i,22}^{-1}(f_i) = -b_i$.

We shall begin, after some preliminaries, with a new presentation of earlier results on the factorization of a special class of structured matrices. Sections 4–7 give a survey of new results and applications.

1.1. Some notation. Let $RH_{p\times q}^\infty$ denote the space of $p \times q$ rational matrix-valued functions $K(z)$ that are analytic and bounded inside the unit disc ($|z| < 1$). A matrix valued function $S(z) \in RH_{p\times q}^\infty$ that is strictly bounded by unity in $|z| < 1$ ($\|S\|_\infty < 1$) will be referred to as a *Schur function*. We also use the notation $\mathcal{H}_A^k(z)$ to refer to the following block-Toeplitz upper-triangular matrix

$$\mathcal{H}_A^k(z) = \begin{bmatrix} A(z) & \frac{1}{1!}A^{(1)}(z) & \frac{1}{2!}A^{(2)}(z) & \cdots & \frac{1}{(k-1)!}A^{(k-1)}(z) \\ & A(z) & \frac{1}{1!}A^{(1)}(z) & \cdots & \frac{1}{(k-2)!}A^{(k-2)}(z) \\ & & A(z) & \cdots & \frac{1}{(k-3)!}A^{(k-3)}(z) \\ & & & & \vdots \\ & \mathbf{O} & & \ddots & \frac{1}{1!}A^{(1)}(z) \\ & & & & A(z) \end{bmatrix},$$

where $A(z)$ is a rational matrix function analytic at z, $k \geq 1$ is a positive integer, and $A^{(i)}(z)$ denotes the ith derivative at z. We denote by $e_i = \begin{bmatrix} 0_{1\times i} & 1 & 0 \end{bmatrix}$ the ith basis vector of the $n-$dimensional space of complex numbers $\mathbf{C}^{1\times n}$. The symbol $*$ stands for Hermitian conjugation (complex conjugation for scalars).

1.2. Basic tools. The mathematical machinery used throughout this work is not much more than elementary matrix and linear systems theory. A key property in our analysis is the simple Gaussian elimination procedure presented now; it will be combined with displacement structure to get a fast factorization algorithm.

We restrict ourselves to Hermitian positive-definite matrices R, even though the results can be extended to more general cases [22,23,24]. A classical algorithm for the triangular factorization of $R = [r_{mj}]_{m,j=0}^{n-1}$ is the so-called Schur reduction procedure. The assumption of positive-definiteness guarantees the existence of a triangular factorization of the form $R = LD^{-1}L^*$, where L is lower-triangular and D is a diagonal matrix with positive entries. The columns of L and the diagonal entries of D can be recursively computed as follows: let l_0 and d_0 denote the first column and the $(0,0)$ entry of R, respectively,

$$d_0 = r_{00}\,, \quad l_0 = \begin{bmatrix} r_{00}\ r_{10}\ \ldots\ r_{n-1,0} \end{bmatrix}^{\mathbf{T}}.$$

If we subtract from R the outer product $l_0 d_0^{-1} l_0^*$ then we obtain a new matrix with one zero row and column. That is,

$$(1.1) \qquad R - l_0 d_0^{-1} l_0^* = \begin{bmatrix} 0 & \mathbf{0} \\ \mathbf{0} & R_1 \end{bmatrix} \equiv \tilde{R}_1\,,$$

where $R_1 = \left[r_{mj}^{(1)} \right]_{m,j=0}^{n-2}$ is called the Schur complement of r_{00} in R. Expression (1.1) represents one Schur reduction step and it can be repeated in order to compute the Schur complement $R_2 = \left[r_{mj}^{(2)} \right]_{m,j=0}^{n-3}$ of $r_{00}^{(1)}$ in R_1, and so on. Each further step corresponds to a recursion of the form

$$(1.2) \qquad \begin{bmatrix} 0 & \mathbf{0} \\ \mathbf{0} & R_{i+1} \end{bmatrix} = R_i - l_i d_i^{-1} l_i^*\,,$$

where $d_i = r_{00}^{(i)}$ (the $(0,0)$ entry of the ith Schur complement R_i), and l_i denotes the first column of R_i. It follows from (1.2) that R can be expressed as the sum of n rank 1 terms (since $R_n = 0$),

$$R = \sum_{i=0}^{n-1} \begin{bmatrix} \mathbf{0}_i \\ l_i \end{bmatrix} d_i^{-1} \begin{bmatrix} \mathbf{0}_i \\ l_i \end{bmatrix}^*.$$

Therefore, $D =$ diagonal $\{d_0, d_1, \ldots, d_{n-1}\}$ and the nonzero parts of the columns of the lower triangular factor L are given by $\{l_i\}_{i=0}^{n-1}$.

In summary, the triangular factors L and D can be constructed from the first columns of the successive Schur complements R_i. Observe however, that the Schur reduction procedure (1.2) is a recursive algorithm that operates directly on the entries of R_i. This requires $O(n^3)$ operations (additions and multiplications). The computational complexity can be reduced to $O(rn^2)$ when R exhibits displacement structure (with displacement rank $r \ll n$), since for structured matrices we can replace (1.2) with an alternative more efficient so-called *generator recursion* to be derived in the next section.

We further state a simple result in matrix theory that plays an important role in the derivation of all so-called square-root algorithms (see, *e.g.*, [25]).

LEMMA 1.1. *Consider two $n \times m$ $(n \leq m)$ matrices A and B. If $AJA^* = BJB^*$ is of full rank, for some $m \times m$ signature matrix J, then there exists a $J-$unitary $m \times m$ matrix Θ ($\Theta J \Theta^* = J$) such that $A = B\Theta$.*

2. Array algorithms. We now consider a positive-definite Hermitian matrix R that has low displacement rank, say r, with respect to the displacement operation $R - FRF^*$. That is, we can write

$$(2.1) \qquad R - FRF^* = GJG^* \ ,$$

for some $n \times r$ so-called generator matrix G and a signature matrix $J = (I_p \oplus -I_q)$. The diagonal entries of the (stable) lower triangular matrix F will be denoted by $\{f_i\}_{i=0}^{n-1}$ ($|f_i| < 1$). The positive-definiteness of R guarantees the existence of a unique (lower triangular) Cholesky factor $\bar{L} = LD^{-1/2}$ such that $R = \bar{L}\bar{L}^*$, and we shall denote the *nonzero* parts of the columns of $\bar{L}$ by $\{\bar{l}_i\}_{i=0}^{n-1}$ ($\bar{l}_i = l_i d_i^{-1/2}$).

Let g_0 denote the first row of G. It follows from the displacement equation (2.1) and from the positive-definiteness of R that

$$d_0 = \frac{g_0 J g_0^*}{1 - |f_0|^2} > 0.$$

Consequently, g_0 has positive $J-$norm ($g_0 J g_0^* > 0$) and we can always choose a $J-$unitary matrix Θ_0 that reduces g_0 to the form (recall Lemma 1.1)

$$(2.2) \qquad g_0 \Theta_0 = \begin{bmatrix} \delta_0 & 0 & \dots & 0 \end{bmatrix} \ ,$$

where δ_0 is a positive scalar. By comparing the $J-$norm on both sides of (2.2) we conclude that the value of δ_0 is given by $\delta_0 = \sqrt{d_0(1 - |f_0|^2)}$. Hence, the action of Θ_0 is to reduce the original generator G to the following form

$$(2.3) \qquad G\Theta_0 = \begin{bmatrix} \delta_0 & 0 & 0 \\ x & x & x \\ x & x & x \\ x & x & x \end{bmatrix} \equiv \bar{G} \ ,$$

where the first row of $\bar{G}$ lies along the direction of the basis vector $[1 \ 0 \ \dots \ 0]$. Clearly, $\bar{G}$ is also a generator of R since $GJG^* = G\Theta_0 J\Theta_0^* G^* = \bar{G}J\bar{G}^*$. We say that $\bar{G}$ is a *proper* generator of R. Moreover, the rotation Θ_0 can be implemented in a variety of ways: by using a sequence of elementary Givens and hyperbolic rotations [26], Householder transformations [27,28], etc..

It further follows from (2.1) that we can write

$$[\bar{L} \ \mathbf{0}] \begin{bmatrix} \bar{L}^* \\ \mathbf{0} \end{bmatrix} = [F\bar{L} \ G] \begin{bmatrix} I & \mathbf{0} \\ \mathbf{0} & J \end{bmatrix} \begin{bmatrix} \bar{L}^* F^* \\ G^* \end{bmatrix} .$$

This last expression fits into the statement of Lemma 1.1. Hence, there exists an $(I \oplus J)-$unitary matrix $\mathbf{\Gamma}$ such that

$$(2.4) \qquad [\bar{L} \ \mathbf{0}] = [F\bar{L} \ G] \, \mathbf{\Gamma}.$$

By examining this identity more closely, we shall derive a fast recursive algorithm for computing the Cholesky factor $\bar{L}$ from knowledge of F and G alone.[1] We first note that the $(I \oplus J)$-unitary transformation Γ can be achieved through a sequence of elementary transformations, say $\Gamma_0, \Gamma_1, \Gamma_2,$ .., that produce the block zero in the postarray by introducing *one zero row at a time*. We first implement Γ_0 as a sequence of two rotations Θ_0 and Γ_0. The first rotation Θ_0 reduces the generator G to proper form, and the second rotation Γ_0 annihilates the remaining nonzero entry δ_0. The overall effect is to annihilate the first row of the G matrix. We then proceed to implement Γ_1, Γ_2, .. in a similar fashion.

So let Θ_0 be a J-unitary matrix that reduces G to proper form, viz.,

$$[\,F\bar{L}\ G\,]\begin{bmatrix} I & \\ & \Theta_0 \end{bmatrix} = [\,F\bar{L}\ \bar{G}\,] \equiv \begin{bmatrix} & \delta_0\ 0\ 0 \\ F\bar{L} & x\ x\ x \\ & x\ x\ x \end{bmatrix}.$$

In order to annihilate the first row of $\bar{G}$ we still need an elementary unitary (Givens) rotation, say Γ_0, that eliminates the nonzero entry δ_0. This can be done by "pivoting" the first column of $\bar{G}$ against the first column of $F\bar{L}$, while keeping all other columns unchanged. This operation produces the first column of $\bar{L}$ (because of (2.4)), and a matrix G_1 whose significance we shall verify very soon:

$$(2.5)\quad \begin{bmatrix} \boxed{f_0 d_0^{1/2}} & \mathbf{0} & \boxed{\delta_0}\ 0\ 0 \\ x & & x\ x\ x \\ x & F_1\bar{L}_1 & x\ x\ x \end{bmatrix} \begin{bmatrix} c & & s \\ & I & \\ s^* & & -c \\ & & I \end{bmatrix} = \begin{bmatrix} \bar{l}_0 & \mathbf{0} & \boxed{0}\ 0\ 0 \\ & F_1\bar{L}_1 & G_1 \end{bmatrix},$$

where F_1 and $\bar{L}_1$ are the submatrices obtained after deleting the first row and column of F and $\bar{L}$, respectively. The letters c and s denote the (cosine and sine) parameters of the rotation matrix. Let $\bar{x}_0$ and x_1 denote the first columns of $\bar{G}$ and G_1, respectively. From expression (2.5) we see that, ignoring the columns that remain unchanged and are thus common to the pre- and post-arrays,

$$(2.6)\qquad [\,F\bar{l}_0\ \bar{x}_0\,]\begin{bmatrix} c & s \\ s^* & -c \end{bmatrix} = \begin{bmatrix} \bar{l}_0 & 0 \\ & x_1 \end{bmatrix},$$

where the top entry of $\bar{x}_0$ is δ_0. The rotation parameters are clearly given by

$$c = \frac{1}{\sqrt{1+|\rho_0|^2}}\ ,\qquad s = \frac{\rho_0}{\sqrt{1+|\rho_0|^2}}\ ,\qquad \rho_0 = \frac{\delta_0}{f_0 d_0^{1/2}}\ .$$

---

[1] This particular approach, one of several possible ones, is a variation of one suggested by H. Lev-Ari (see [31, Chapter 2] and also Section 3 ahead).

That is, we can rewrite (2.6) more explicitly as follows:

$$\begin{bmatrix} F\bar{l}_0 & \bar{x}_0 \end{bmatrix} \begin{bmatrix} f_0^* & \frac{\delta_0}{d_0^{1/2}} \\ \frac{\delta_0}{d_0^{1/2}} & -f_0 \end{bmatrix} = \begin{bmatrix} \bar{l}_0 & 0 \\ & x_1 \end{bmatrix} \;,$$

which leads to

$$\begin{bmatrix} 0 \\ x_1 \end{bmatrix} = \Phi_0 \bar{x}_0 = \Phi_0 G \Theta_0 \begin{bmatrix} 1 \\ 0 \end{bmatrix} \;,$$

where we defined the "Blaschke" matrix $\Phi_0 = (I - f_0^* F)^{-1}(F - f_0 I)$. We still need to verify the significance of G_1. Comparing the $(I \oplus J)$-norm on both sides of (2.5) we obtain

$$F \bar{L}\bar{L}^* F^* + \bar{G} J \bar{G}^* = \bar{l}_0 \bar{l}_0^* + \begin{bmatrix} 0 \\ F_1 \bar{L}_1 \end{bmatrix} \begin{bmatrix} 0 & \bar{L}_1^* F_1^* \end{bmatrix} + \begin{bmatrix} 0 \\ G_1 \end{bmatrix} J \begin{bmatrix} 0 & G_1^* \end{bmatrix} .$$

But the Cholesky factor of the first Schur complement R_1 is $\bar{L}_1$ itself. Hence, using (2.1) we get

$$R - \bar{l}_0 \bar{l}_0^* = \begin{bmatrix} 0 & 0 \\ 0 & F_1 R_1 F_1^* \end{bmatrix} + \begin{bmatrix} 0 & 0 \\ 0 & G_1 J G_1^* \end{bmatrix} .$$

Consequently (recall (1.1)), $R_1 - F_1 R_1 F_1^* = G_1 J G_1^*$, which shows that G_1 is a generator matrix of the Schur complement R_1 with respect to the displacement operation $R_1 - F_1 R_1 F_1^*$. Hence, G_1 is obtained as follows: choose a J-unitary rotation Θ_0 that converts the first row of G to the form $\begin{bmatrix} \delta_0 & 0 \end{bmatrix}^{\mathbf{T}}$ and apply Θ_0 to G as in (2.3); keep the last $(r - 1)$ columns of $G\Theta_0$ unchanged and multiply the first column by Φ_0; this results in G_1. We can write this transformation in the following compact (array) form:

$$\begin{bmatrix} 0 \\ G_1 \end{bmatrix} = G \Theta_0 \begin{bmatrix} 0 & 0 \\ 0 & I_{r-1} \end{bmatrix} + \Phi_0 G \Theta_0 \begin{bmatrix} 1 & 0 \\ 0 & 0 \end{bmatrix} .$$

Therefore, the effect of the transformation $\mathbf{\Gamma}_0$ (which we implemented as a sequence of two rotations Θ_0 and Γ_0) is to annihilate the first row of G,

$$\begin{bmatrix} F\bar{L} & G \end{bmatrix} \mathbf{\Gamma}_0 = \begin{bmatrix} \bar{l}_0 & 0 & 0 \\ & F_1 \bar{L}_1 & G_1 \end{bmatrix} .$$

We can now proceed by annihilating the first row of G_1,

$$\begin{bmatrix} F_1 \bar{L}_1 & G_1 \end{bmatrix} \mathbf{\Gamma}_1 = \begin{bmatrix} \bar{l}_1 & 0 & 0 \\ & F_2 \bar{L}_2 & G_2 \end{bmatrix} \;,$$

where F_2 and $\bar{L}_2$ are the submatrices obtained after deleting the first row and column of F_1 and $\bar{L}_1$, respectively, and so on.

Moreover, it follows from the displacement equation (2.1) that the first column of R satisfies the relation $l_0 = Fl_0f_0^* + GJg_0^*$. Hence, using the properness of $G\Theta_0$ (see (2.3)) we have

$$l_0 = (I - f_0^*F)^{-1}G\Theta_0 J\Theta_0^*g_0^* = (I - f_0^*F)^{-1}\bar{x}_0\delta_0.$$

Using the fact that $\bar{l}_0 = l_0 d_0^{-1/2}$, we get

$$(2.7) \qquad \bar{l}_0 = \sqrt{1 - |f_0|^2}\,(I - f_0^*F)^{-1}\bar{x}_0\,, \qquad \text{since} \quad d_0 = \frac{\delta_0^2}{1 - |f_0|^2}.$$

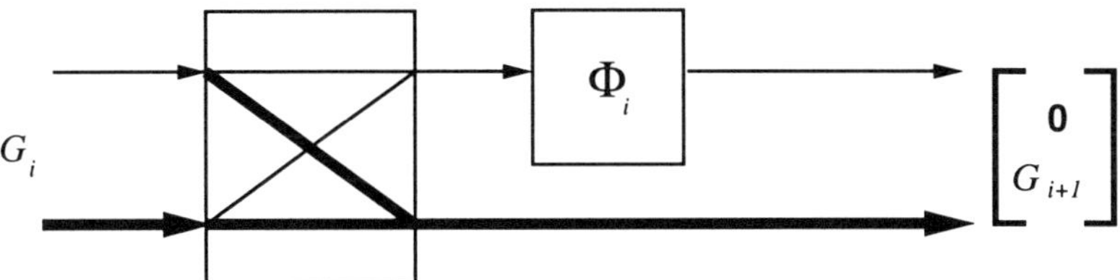

FIG. 2.1. *One step of the generator recursion.*

Similar expressions are valid for the other column vectors $\bar{l}_i$, $i \geq 1$. In summary, we are led to the following recursive procedure (see [29,32] for earlier and different derivations).

Algorithm 2.1. The Cholesky factorization of a positive-definite Hermitian matrix R with displacement structure of the form $R - FRF^* = GJG^*$, can be computed by the recursive procedure

$$(2.8) \qquad \begin{bmatrix} \mathbf{0} \\ G_{i+1} \end{bmatrix} = G_i\Theta_i \begin{bmatrix} 0 & 0 \\ 0 & I_{r-1} \end{bmatrix} + \Phi_i G_i\Theta_i \begin{bmatrix} 1 & 0 \\ 0 & 0 \end{bmatrix}, \qquad F_0 = F,\ \ G_0 = G\ ,$$

$$\Phi_i = (I_{n-i} - f_i^*F_i)^{-1}(F_i - f_iI_{n-i})\ ,$$

where F_i is the submatrix obtained after deleting the first row and column of F_{i-1}, and Θ_i is an arbitrary J-unitary matrix that reduces the first row of G_i (denoted by g_i) to the form $g_i\Theta_i = \begin{bmatrix} \delta_i & 0 \ldots 0 \end{bmatrix}$. The columns of the Cholesky factor $\bar{L}$ are then given by

$$\bar{l}_i = \sqrt{1 - |f_i|^2}\,(I_{n-i} - f_i^*F_i)^{-1}G_i\Theta_i \begin{bmatrix} 1 \\ 0 \end{bmatrix}.$$

Pictorially, we have the following simple array picture as depicted in Figure 2.1.

$$G_i = \begin{bmatrix} x & x & x \\ x & x & x \\ x & x & x \\ \vdots & \vdots & \vdots \end{bmatrix} \xrightarrow{\Theta_i} \begin{bmatrix} \delta_i & 0 & 0 \\ x & x & x \\ x & x & x \\ \vdots & \vdots & \vdots \end{bmatrix} \xrightarrow{\Phi_i} \begin{bmatrix} 0 & 0 & 0 \\ x & x & x \\ x & x & x \\ \vdots & \vdots & \vdots \end{bmatrix} = \begin{bmatrix} \mathbf{0} \\ G_{i+1} \end{bmatrix}.$$

2.1. First-order J-lossless sections. It follows from the square-root argument (using (2.5)) that the expressions for l_i and G_i can be grouped together into the following revealing expression:

$$(2.9) \qquad \begin{bmatrix} l_i & 0 \\ & G_{i+1} \end{bmatrix} = \begin{bmatrix} F_i l_i & G_i \end{bmatrix} \begin{bmatrix} f_i^* & \frac{\delta_i}{d_i}\begin{bmatrix} 1 & 0 \end{bmatrix} \\ \Theta_i \begin{bmatrix} \delta_i \\ 0 \end{bmatrix} & \Theta_i \begin{bmatrix} -f_i & 0 \\ 0 & I_{r-1} \end{bmatrix} \end{bmatrix},$$

which clearly shows that each step of the generator recursion involves a first-order state-space system that appears on the right-hand-side of the above expression. Let $\Theta_i(z)$ denote its $r \times r$ transfer matrix (with inputs from the left), viz.,

$$\Theta_i(z) = \Theta_i \begin{bmatrix} -f_i & 0 \\ 0 & I_{r-1} \end{bmatrix} + \Theta_i \begin{bmatrix} \delta_i \\ 0 \end{bmatrix} (z^{-1} - f_i^*)^{-1} \frac{\delta_i}{d_i} \begin{bmatrix} 1 & 0 \end{bmatrix}.$$

It then readily follows, upon simplification, that

$$(2.10) \qquad \Theta_i(z) = \Theta_i \begin{bmatrix} \frac{z-f_i}{1-zf_i^*} & 0 \\ 0 & I_{r-1} \end{bmatrix}.$$

Each such section is clearly J-lossless. This follows from the fact that $\Theta_i(z)$ is analytic in $|z| < 1$ due to $|f_i| < 1$, and that $\Theta_i(z)J\Theta_i^*(z) = J$ on $|z| = 1$ since $(z - f_i)/(1 - zf_i^*)$ is a Blaschke factor and Θ_i is J-unitary. Furthermore, each $\Theta_i(z)$ also has an important "blocking" property that will be very relevant in the solution of interpolation problems.

LEMMA 2.2. *Each first-order section $\Theta_i(z)$ has a transmission zero at f_i and along the direction defined by g_i, viz., $g_i\Theta_i(f_i) = 0$.*

Proof. This is evident from the relation

$$g_i\Theta_i(f_i) \quad = \quad g_i\Theta_i \begin{bmatrix} 0 & 0 \\ 0 & I_{r-1} \end{bmatrix} = \begin{bmatrix} \delta_i & 0 \end{bmatrix} \begin{bmatrix} 0 & 0 \\ 0 & I_{r-1} \end{bmatrix} = 0.$$

$\square$

3. General algorithm. The algorithm derived in Section 2 is in a convenient array form. We verify here that it is a special case of a more general recursion, which under suitable manipulations reduces to the array form discussed above.

THEOREM 3.1. *The Schur complements R_i are also structured with generator matrices G_i, viz., $R_i - F_i R_i F_i^* = G_i J G_i^*$, where G_i is an $(n-i) \times r$ generator matrix that satisfies, along with l_i, the following recursion*

$$(3.1) \qquad \begin{bmatrix} l_i & 0 \\ & G_{i+1} \end{bmatrix} = \begin{bmatrix} F_i l_i & G_i \end{bmatrix} \begin{bmatrix} f_i^* & h_i^* J \\ J g_i^* & J k_i^* J \end{bmatrix},$$

where g_i is the first row of G_i, and h_i and k_i are arbitrary $r \times 1$ and $r \times r$ matrices, respectively, chosen so as to satisfy the embedding relation

$$(3.2) \qquad \begin{bmatrix} f_i & g_i \\ h_i & k_i \end{bmatrix} \begin{bmatrix} d_i & 0 \\ 0 & J \end{bmatrix} \begin{bmatrix} f_i & g_i \\ h_i & k_i \end{bmatrix}^* = \begin{bmatrix} d_i & 0 \\ 0 & J \end{bmatrix} ,$$

where

$$(3.3) \qquad d_i = \frac{g_i J g_i^*}{1 - |f_i|^2} ,$$

and F_i is the $(n-i) \times (n-i)$ submatrix obtained after deleting the first row and column of F_{i-1}.

Proof. We prove the result for $i = 0$. The same argument holds for $i \geq 1$. Using (1.1) we write

$$\tilde{R}_1 - F\tilde{R}_1 F^* = -\frac{1}{d_0}\left[f_0^* F l_0 g_0 J G^* + G J g_0^* l_0^* F^* f_0 - \frac{1}{d_0} F l_0 g_0 J g_0^* l_0^* F^* \right] +$$

$$GJ\left\{ J - \frac{g_0^* g_0}{d_0} \right\} J G^*.$$

(3.4)

We now verify that the right-hand side of the above expression can be put into the form of a *perfect square* by introducing some auxiliary quantities. Consider an $r \times 1$ column vector h_0 and an $r \times r$ matrix k_0 that are defined to satisfy the following relations (in terms of the quantities that appear on the right-hand side of the above expression. We shall see very soon that this is always possible):

$$(3.5) \quad h_0^* J h_0 = \frac{g_0 J g_0^*}{d_0^2} , \qquad k_0^* J k_0 = J - \frac{g_0^* g_0}{d_0} , \qquad k_0^* J h_0 = -\frac{f_0 g_0^*}{d_0} .$$

Using $\{h_0, k_0\}$, we can rewrite the right-hand side of (3.4) in the form

$$GJ k_0^* J k_0 J G^* + GJ k_0^* J h_0 l_0^* F^* + F l_0 h_0^* J k_0 J G^* + F l_0 h_0^* J h_0 l_0^* F^*,$$

which can clearly be factored as $\tilde{G}_1 J \tilde{G}_1^*$, where $\tilde{G}_1 = F l_0 h_0^* J + GJ k_0^* J$. But the first row and column of $\tilde{R}_1$ are zero. Hence, the first row of $\tilde{G}_1$ is zero, $\tilde{G}_1 = \begin{bmatrix} 0 & G_1^{\mathrm{T}} \end{bmatrix}^{\mathrm{T}}$. Moreover, it follows from (3.5) (and the expression for d_0) that $\{f_0, g_0, h_0, k_0\}$ satisfy the relation

$$\begin{bmatrix} f_0 & g_0 \\ h_0 & k_0 \end{bmatrix}^* \begin{bmatrix} d_0^{-1} & 0 \\ 0 & J \end{bmatrix} \begin{bmatrix} f_0 & g_0 \\ h_0 & k_0 \end{bmatrix} = \begin{bmatrix} d_0^{-1} & 0 \\ 0 & J \end{bmatrix} ,$$

which is equivalent to (3.2) for $i = 0$. $\qquad \square$

It is worth noting that the generator recursion (3.1) has the same form as the array equation (2.9) that we wrote earlier. In fact, the matrix defined by

$$\begin{bmatrix} f_i^* & h_i^* J \\ J g_i^* & J k_i^* J \end{bmatrix}$$

is the general form of an elementary transformation that produces the desired zero row on the left-hand side of (3.1). Moreover, if we consider the transfer matrix $\Theta_i(z)$ associated with the above discrete-time system, viz.,

$$(3.6) \qquad \Theta_i(z) \quad = \quad J k_i^* J + J g_i^* \left[z^{-1} - f_i^* \right]^{-1} h_i^* J.$$

Then, using the embedding relation (3.2) (or the expressions similar to (3.5) for h_i and k_i), we readily conclude that

$$(3.7) \qquad \Theta_i(z) J \Theta_i^*(z) = J + \frac{J g_i^* g_i J}{d_i} \frac{z z^* - 1}{(1 - z f_i^*)(1 - z^* f_i)} ,$$

which confirms that each first-order section $\Theta_i(z)$ is J-lossless. Furthermore, the blocking property of $\Theta_i(z)$ is also evident here since

$$g_i \Theta_i(f_i) = g_i J k_i^* J + g_i J g_i^* \frac{f_i}{1 - |f_i|^2} h_i^* J = g_i J k_i^* J + f_i d_i h_i^* J \overset{(3.2)}{=} \mathbf{0}.$$

Using the embedding relation (3.2) we can further show [29,30,31], following an argument similar to that in [29], that all choices of h_i and k_i are completely specified by $\{f_i, g_i, d_i\}$.

LEMMA 3.2. *All possible choices of h_i and k_i are given by*

$$h_i = \Theta_i^{-1} \left\{ \frac{1}{d_i} \frac{\tau_i - f_i}{1 - \tau_i f_i^*} J g_i^* \right\} \quad and \quad k_i = \Theta_i^{-1} \left\{ I_r - \frac{1}{d_i} \frac{J g_i^* g_i}{1 - \tau_i f_i^*} \right\} ,$$

(3.8)
for an arbitrary J-unitary matrix Θ_i and an arbitrary scalar τ_i on the unit circle $(|\tau_i| = 1)$.

Using expression (3.8) for h_i and k_i we can rewrite the generator recursion (3.1) and the transfer matrix (3.6) in a more convenient form that depends (up to J-unitary rotations) only on known parameters (see also [29,32] for earlier and alternative derivations).

THEOREM 3.3. *The generator recursion (3.1) and the transfer matrix (3.6) reduce to*

$$(3.9) \qquad \begin{bmatrix} \mathbf{0} \\ G_{i+1} \end{bmatrix} = \left\{ G_i + (\Phi_i - I_{n-i}) G_i \frac{J g_i^* g_i}{g_i J g_i^*} \right\} \Theta_i ,$$

$$(3.10) \qquad \Theta_i(z) = \left\{ I_r + [B_i(z) - 1] \frac{J g_i^* g_i}{g_i J g_i^*} \right\} \Theta_i ,$$

where $B_i(z)$ is a Blaschke factor of the form

$$B_i(z) = \frac{z - f_i}{1 - z f_i^*} \frac{1 - \tau_i f_i^*}{\tau_i - f_i} ,$$

and Φ_i is a "Blaschke" matrix given by,

$$\Phi_i = \frac{1 - \tau_i f_i^*}{\tau_i - f_i}(F_i - f_i I_{n-i})(I_{n-i} - f_i^* F_i)^{-1}.$$

We remark that $(F_i - f_i I_{n-i})$ and $(I_{n-i} - f_i^* F_i)^{-1}$ commute, and hence the expression for Φ_i given above will be the same as the expression given earlier in Algorithm 2 if we choose

$$(3.11) \qquad \tau_i = \frac{1 + f_i}{1 + f_i^*}.$$

Notice also that the blocking property of each Section $\Theta_i(z)$ is again evident from expression (3.10) since $B_i(f_i) = 0$. That is,

$$g_i \Theta_i(f_i) = \left\{ g_i - \frac{g_i J g_i^*}{g_i J g_i^*} \, g_i \right\} \Theta_i = 0.$$

The generator recursion of Theorem 3.3 is the general form of the factorization algorithm and it includes, as special cases, the array algorithm derived in Section 2. Observe for instance, that (3.9) has two parameters that we are free to choose: Θ_i and τ_i. Choosing τ_i as in (3.11) and Θ_i such that $g_i \Theta_i$ is reduced to the form in (2.2), we can easily check that Theorem 3.3 reduces to the array algorithm of Section 2.

4. The tangential Hermite-Fejér problem. We now show that the algorithm derived in the previous sections also solves interpolation problems. We first state a general Hermite-Fejér interpolation problem that includes many of the classical problems as special cases. We consider m points $\{\alpha_i\}_{i=0}^{m-1}$ inside the open unit disc $\mathbf{D}$ and we associate with each point α_i a positive integer $r_i \geq 1$ and two row vectors $\mathbf{a}_i$ and $\mathbf{b}_i$ partitioned as follows:

$$\mathbf{a}_i = \left[u_1^{(i)} \ u_2^{(i)} \ \ldots \ u_{r_i}^{(i)} \right], \quad \mathbf{b}_i = \left[v_1^{(i)} \ v_2^{(i)} \ \ldots \ v_{r_i}^{(i)} \right],$$

where $u_j^{(i)}$ and $v_j^{(i)}$ $(j = 1, \ldots, r_i)$ are $1 \times p$ and $1 \times q$ row vectors, respectively. That is, $\mathbf{a}_i$ and $\mathbf{b}_i$ are partitioned into r_i row vectors each. If an interpolating point α_i is repeated (say, $\alpha_i = \alpha_{i+1} = \ldots = \alpha_{i+j}$), then we shall further assume that the following condition is satisfied (which rules out degenerate cases [30]):

$$(4.1) \qquad \{u_1^{(i)}, u_1^{(i+1)}, \ldots, u_1^{(i+j)}\} \quad \text{are linearly independent.}$$

The tangential Hermite-Fejér problem then reads as follows (see, *e.g.*, [5]).

Problem 4.1. Describe all Schur-type functions $S(z) \in RH^{\infty}_{p \times q}$ that satisfy

$$(4.2) \qquad \mathbf{b}_i = \mathbf{a}_i \mathcal{H}^{r_i}_S(\alpha_i) \quad \text{for} \quad 0 \leq i \leq m - 1.$$

This statement clearly includes, as special cases, the problems of Carathéodory-Fejér [33,34,35], Nevanlinna-Pick [33,36,37], and the corresponding tangential (matrix) versions.

4.1. Solvability condition. The first step in the recursive solution consists in constructing three matrices F, G, and J directly from the interpolation data: F contains the information relative to the points $\{\alpha_i\}$ and the dimensions $\{r_i\}$, G contains the information relative to the direction vectors $\{\mathbf{a}_i\}$ and $\{\mathbf{b}_i\}$, and $J = (I_p \oplus -I_q)$ is a signature matrix. The matrices F and G are constructed as follows: we associate with each α_i a Jordan block $\bar{F}_i$ of size $r_i \times r_i$,

$$\bar{F}_i = \begin{bmatrix} \alpha_i & & & \\ 1 & \alpha_i & & \\ & \ddots & \ddots & \\ & & 1 & \alpha_i \end{bmatrix},$$

and two $r_i \times p$ and $r_i \times q$ matrices U_i and V_i, respectively, which are composed of the row vectors associated with α_i, viz.,

$$U_i = \begin{bmatrix} u_1^{(i)} \\ u_2^{(i)} \\ \vdots \\ u_{r_i}^{(i)} \end{bmatrix} \quad \text{and} \quad V_i = \begin{bmatrix} v_1^{(i)} \\ v_2^{(i)} \\ \vdots \\ v_{r_i}^{(i)} \end{bmatrix}.$$

Then $F = $ diagonal $\{\bar{F}_0, \bar{F}_1, \ldots, \bar{F}_{m-1}\}$ and

$$(4.3) \qquad G = \begin{bmatrix} U_0 & V_0 \\ U_1 & V_1 \\ \vdots & \vdots \\ U_{m-1} & V_{m-1} \end{bmatrix} \equiv [\, \mathbf{U} \, \mathbf{V} \,].$$

Let $n = \sum_{i=0}^{m-1} r_i$ and $r = p + q$, then F and G are $n \times n$ and $n \times r$ matrices, respectively. We shall denote the diagonal entries of F by $\{f_i\}_{i=0}^{n-1}$ (for example, $f_0 = f_1 = \ldots = f_{r_0 - 1} = \alpha_0$). We also associate with the interpolation Problem 4 the following displacement equation

$$(4.4) \qquad R - FRF^* = GJG^*.$$

R is clearly unique since F is a stable matrix ($|f_i| < 1,\ \forall\, i$). We shall prove in the next section that by applying the array algorithm to F and G we obtain a transmission-line cascade $\Theta(z)$ that parametrizes all solutions of the Hermite-Fejér problem. Meanwhile, we verify that the above construction of F, G, and R allows us to prove the necessary and sufficient conditions for the existence of solutions (see also [38,39] for related discussion).

THEOREM 4.2. *The tangential Hermite-Fejér problem is solvable if, and only if, R is positive-definite.*

Proof. If R is positive-definite then the recursive procedure described later finds a solution $S(z)$. Conversely, assume there exists a solution $S(z)$ satisfying the interpolation conditions (4.2), and let $\{S_i\}_{i=0}^{\infty}$ be the Taylor series coefficients of $S(z)$ around the origin, viz.,

$$S(z) = S_0 + zS_1 + z^2 S_2 + z^3 S_3 + \ldots$$

Define the (semi-infinite) block lower-triangular Toeplitz matrix

$$\mathcal{S} = \begin{bmatrix} S_0 & & & \\ S_1 & S_0 & & \mathbf{O} \\ S_2 & S_1 & S_0 & \\ \vdots & & & \ddots \end{bmatrix},$$

as well as the (semi-infinite) matrices

$$\mathcal{U} = \begin{bmatrix} \mathbf{U} & F\mathbf{U} & F^2\mathbf{U} & \ldots \end{bmatrix} \quad \text{and} \quad \mathcal{V} = \begin{bmatrix} \mathbf{V} & F\mathbf{V} & F^2\mathbf{V} & \ldots \end{bmatrix}.$$

We can easily check that because of (4.2) we get $\mathcal{V} = \mathcal{U}\mathcal{S}$. But R in (4.4) is given by

$$R = \mathcal{U}\mathcal{U}^* - \mathcal{V}\mathcal{V}^* = \mathcal{U}\left(I - \mathcal{S}\mathcal{S}^*\right)\mathcal{U}^*.$$

Moreover, $\mathcal{S}$ is a strict contraction (since $S(z)$ is a Schur-type function with $\|S\|_{\infty} < 1$) and it follows from (4.1) that $\mathcal{U}\mathcal{U}^* > 0$ (see [30,31]). Hence, $R > 0$. $\square$

4.2. Interpolation properties. We already know how to construct a convenient structure (4.4) from the interpolation data. We remark that we only know F, G, and J, whereas the matrix R itself *is not known* a priori. In fact, the recursive procedure described here *does not* require R. It only uses the matrices F, G, and J that are constructed *directly* from the interpolation data.

We now verify that if we apply the array algorithm to G in (4.4), we then obtain a cascade $\Theta(z)$,

$$\Theta(z) = \Theta_0(z)\Theta_1(z)\ldots\Theta_{n-1}(z),$$

of first order J-lossless sections that parametrizes all solutions of the Hermite-Fejér interpolation problem. This follows from the fact that the first-order sections have local blocking properties, $g_i \Theta_i(f_i) = 0$, which reflect into a global blocking property for the entire cascade, as we readily verify.

Consider the first-order section $\Theta_0(z)$. It follows from its local blocking property that

$$e_0 G \Theta_0(f_0) = g_0 \Theta_0(f_0) = 0$$

But the Jordan structure of $\bar{F}_0$ (with eigenvalue $\alpha_0 = f_0 = f_1 = \ldots = f_{r_0-1}$) imposes a stronger condition on $\Theta_0(z)$. Note for example, that the following relation follows immediately from the array form (2.8) (g_0 and g_1 are the first rows of G and G_1, respectively)

$$g_1 = g_0 \Theta_0^{(1)}(f_0) + e_1 G \Theta_0(f_0) \ .$$

More precisely, by comparing the second row on both sides of (2.8) for $i = 0$, we conclude that

$$\begin{aligned}
g_1 &= e_1 G \Theta_0 \begin{bmatrix} 0 & 0 \\ 0 & I_{r-1} \end{bmatrix} + e_1 \Phi_0 G \Theta_0 \begin{bmatrix} 1 & 0 \\ 0 & 0 \end{bmatrix} \\
&= e_1 G \Theta_0(f_0) + \frac{1}{1 - |f_0|^2} e_0 G \Theta_0 \begin{bmatrix} 1 & 0 \\ 0 & 0 \end{bmatrix} \\
&= e_1 G \Theta_0(f_0) + g_0 \Theta_0^{(1)}(f_0).
\end{aligned}$$

Therefore, the first row of G_1 is obtained as a linear combination of the first two rows of G,

$$g_1 \equiv e_0 G_1 = \begin{bmatrix} e_0 G & e_1 G \end{bmatrix} \begin{bmatrix} \Theta_0^{(1)}(f_0) \\ \Theta_0(f_0) \end{bmatrix} .$$

This result can be extended to show that the kth row of G_1 ($k < r_0$) is obtained as a linear combination of the first $(k + 1)$ rows of G, and so on. Putting these remarks together leads to

$$(4.5) \qquad \begin{bmatrix} e_0 G & e_1 G & \ldots & e_{r_0-1} G \end{bmatrix} \mathcal{H}_{\Theta_0}^{r_0}(\alpha_0) = \begin{bmatrix} 0 & e_0 G_1 & e_1 G_1 & \ldots & e_{r_0-2} G_1 \end{bmatrix} .$$

Therefore, when the first r_0 rows of G propagate through $\Theta_0(z)$ we obtain the first $r_0 - 1$ rows of G_1 at $z = \alpha_0$. This argument can be continued [30,31] to conclude the following result: let s_i denote the total size of the Jordan blocks prior to $\bar{F}_i$: $s_i = \sum_{p=0}^{i-1} r_p$, $s_0 = 0$.

THEOREM 4.3. *The transfer matrix $\Theta(z)$ satisfies the global blocking property*

$$(4.6) \qquad \begin{bmatrix} e_{s_i} G & e_{s_i+1} G & \ldots & e_{s_i+r_i-1} G \end{bmatrix} \mathcal{H}_{\Theta}^{r_i}(\alpha_i) = 0.$$

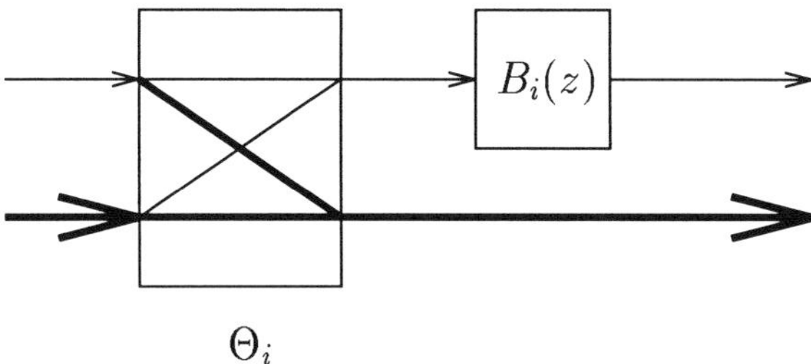

FIG. 4.1. *A J-lossless first-order section* $\Theta_i(z)$.

The row vector on the left hand-side of (4.6) is composed of the r_i row vectors in $\begin{bmatrix} U_i & V_i \end{bmatrix}$ associated with α_i, viz., $\begin{bmatrix} u_1^{(i)} & v_1^{(i)} & u_2^{(i)} & v_2^{(i)} & \dots & u_{r_i}^{(i)} & v_{r_i}^{(i)} \end{bmatrix}$. If we now partition $\Theta(z)$ accordingly with $J = (I_p \oplus -I_q)$,

$$\Theta(z) = \begin{bmatrix} \Theta_{11}(z) & \Theta_{12}(z) \\ \Theta_{21}(z) & \Theta_{22}(z) \end{bmatrix},$$

it is then a standard result that $S(z) = -\Theta_{12}(z)\Theta_{22}^{-1}(z)$ is a Schur-type function due to the J-losslessness of $\Theta(z)$, and we conclude from (4.6) that it satisfies the required interpolation conditions. Moreover, all solutions $S(z)$ are parametrized in terms of a linear fractional transformation based on $\Theta(z)$ (see [11,13,21,30] for details and related discussion).

LEMMA 4.4. *All solutions* $S(z)$ *of the tangential Hermite-Fejér problem are given by a linear fractional transformation of a Schur matrix function* $K(z)$ *(*$\|K\|_\infty < 1$*)*

$$(4.7) \quad S(z) = -\left[\Theta_{11}(z)K(z) + \Theta_{12}(z)\right]\left[\Theta_{21}(z)K(z) + \Theta_{22}(z)\right]^{-1}.$$

4.3. Transmission-line structure. Each section $\Theta_i(z)$ can be schematically represented as shown in Figure 4.1. Figure 4.2 shows a scattering interpretation of the cascade $\Theta(z)$, where $\Sigma(z)$ is the scattering matrix defined by

$$\Sigma(z) = \begin{bmatrix} \Theta_{11} - \Theta_{12}\Theta_{22}^{-1}\Theta_{21} & -\Theta_{12}\Theta_{22}^{-1} \\ \Theta_{22}^{-1}\Theta_{21} & \Theta_{22}^{-1} \end{bmatrix}(z).$$

The solution $S(z)$ is the transfer matrix from the top left $(1 \times p)$ input to the bottom left $(1 \times q)$ output, with a Schur-type load $(-K(z))$ at the right end. Therefore, we are led to the following $O(rn^2)$ recursive algorithm for the solution of the Hermite-Fejér problem.

Algorithm 4.5. The Hermite-Fejér problem can be recursively solved as follows:
- Construct F, G, and J from the interpolation data as described in Section 4.1.

- Start with $F_0 = F$, $G_0 = G$, and apply the array form (2.8) of the generator recursion for $i = 0, 1, \ldots, n-1$.
- Each step provides a first-order section $\Theta_i(z)$ completely specified by f_i, g_i, and Θ_i as in (2.10) or (3.10).
- The cascade of sections $\Theta(z)$ satisfies the relation

$$\left[u_1^{(i)} \; v_1^{(i)} \; u_2^{(i)} \; v_2^{(i)} \; \ldots \; u_{r_i}^{(i)} \; v_{r_i}^{(i)} \right] \mathcal{H}_\Theta^{r_i}(\alpha_i) = \mathbf{0}, \quad 0 \le i \le m-1.$$

- Then $S(z) = -\Theta_{12}(z)\Theta_{22}^{-1}(z)$ satisfies

$$\mathbf{b}_i = \mathbf{a}_i \mathcal{H}_S^{r_i}(\alpha_i), \quad 0 \le i \le m-1.$$

- All solutions $S(z)$ are parametrized by an arbitrary Schur function $K(z)$,

$$S(z) = -\left[\Theta_{11}(z)K(z) + \Theta_{12}(z)\right]\left[\Theta_{21}(z)K(z) + \Theta_{22}(z)\right]^{-1}.$$

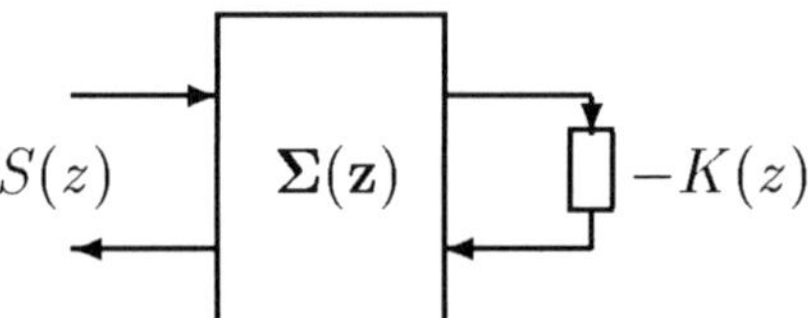

FIG. 4.2. *Scattering interpretation.*

5. Time-variant displacement structure. We now extend the notion of displacement structure to the time-variant setting and show that we can also study matrices that exhibit structured time-variations, special cases of which often arise in adaptive filtering [40], time-variant interpolation [20], and matrix completion problems [21].

We consider an $n \times n$ time-variant positive-definite Hermitian matrix $R(t) = [r_{mj}(t)]_{m,j=0}^{n-1}$, and we shall say that it has a *time-variant Toeplitz-like* structure if the difference,

$$R(t) - F(t)R(t-1)F^*(t),$$

has low rank, say $r(t)$ (usually $r(t) \ll n$), for some lower triangular $n \times n$ matrix $F(t)$ whose diagonal entries we denote by $\{f_i(t)\}_{i=0}^{n-1}$ ($|f_i(t)| < 1$). It follows from the low rank property that we can write

$$(5.1) \qquad R(t) - F(t)R(t-1)F^*(t) = G(t)J(t)G^*(t) \;,$$

where $G(t)$ is an $n \times r(t)$ so-called generator matrix, and $J(t) = (I_{p(t)} \oplus -I_{q(t)})$ is an $r(t) \times r(t)$ signature matrix. The main question that we treat in this section is the following: given the Cholesky factor of $R(t-1)$ and knowing that $R(t)$ satisfies a displacement equation of the form (5.1), how to efficiently and recursively determine the Cholesky factor of $R(t)$?

We may repeat here the same square-root argument as is in the time-invariant case (Section 2) [40,31]. We shall instead, present the general algorithm and discuss applications in time-variant interpolation and completion problems.

Following the same reasoning as in Section 3 we can prove the following result.

THEOREM 5.1. *The Schur complements $R_i(t)$ are also structured with generator matrices $G_i(t)$, viz.,*

$$R_i(t) - F_i(t)R_i(t-1)F_i^*(t) = G_i(t)J(t)G_i^*(t) \ ,$$

where $G_i(t)$ is an $(n-i) \times r(t)$ generator matrix that satisfies, along with $l_i(t)$, the following recursion

$$(5.2) \quad \begin{bmatrix} l_i(t) & \mathbf{0} \\ & G_{i+1}(t) \end{bmatrix} = \begin{bmatrix} F_i(t)l_i(t-1) & G_i(t) \end{bmatrix} \begin{bmatrix} f_i^*(t) & h_i^*(t)J(t) \\ J(t)g_i^*(t) & J(t)k_i^*(t)J(t) \end{bmatrix} \ ,$$

where $g_i(t)$ is the first row of $G_i(t)$, and $h_i(t)$ and $k_i(t)$ are arbitrary $r(t) \times 1$ and $r(t) \times r(t)$ matrices, respectively, chosen so as to satisfy the embedding relation

$$(5.3) \quad \begin{bmatrix} f_i(t) & g_i(t) \\ h_i(t) & k_i(t) \end{bmatrix} \begin{bmatrix} d_i(t-1) & \mathbf{0} \\ \mathbf{0} & J(t) \end{bmatrix} \begin{bmatrix} f_i(t) & g_i(t) \\ h_i(t) & k_i(t) \end{bmatrix}^* = \begin{bmatrix} d_i(t) & \mathbf{0} \\ \mathbf{0} & J(t) \end{bmatrix} \ ,$$

where

$$(5.4) \qquad d_i(t) = |f_i(t)|^2 d_i(t-1) + g_i(t)J(t)g_i^*(t) \ ,$$

and $F_i(t)$ is the $(n-i) \times (n-i)$ submatrix obtained after deleting the first row and column of $F_{i-1}(t)$.

The generator recursion (5.2) has a transmission-line picture in terms of a cascade of elementary sections as shown in Figure 5.1, where each section depends on the parameters $\{f_i(t), g_i(t), h_i(t), k_i(t)\}$. The Δ block represents a storage element where the present value of $l_i(t)$ is stored for the next time instant, and the block with $F_i(t)$ can be implemented as a tapped-delay filter with time-variant coefficients [20,40,31].

LEMMA 5.2. *All possible choices of $h_i(t)$ and $k_i(t)$ are given by*

$$h_i(t) = \Theta_i^{-1}(t) \left\{ \frac{1 - \tau_i^*(t)f_i(t)}{\tau_i^*(t)d_i(t) - d_i(t-1)f_i^*(t)} \ J(t)g_i^*(t) \right\} \ ,$$

$$(5.5) \qquad k_i(t) = \Theta_i^{-1}(t) \left\{ I_{r(t)} - \frac{\tau_i^*(t)J(t)g_i^*(t)g_i(t)}{\tau_i^*(t)d_i(t) - d_i(t-1)f_i^*(t)} \right\} \ ,$$

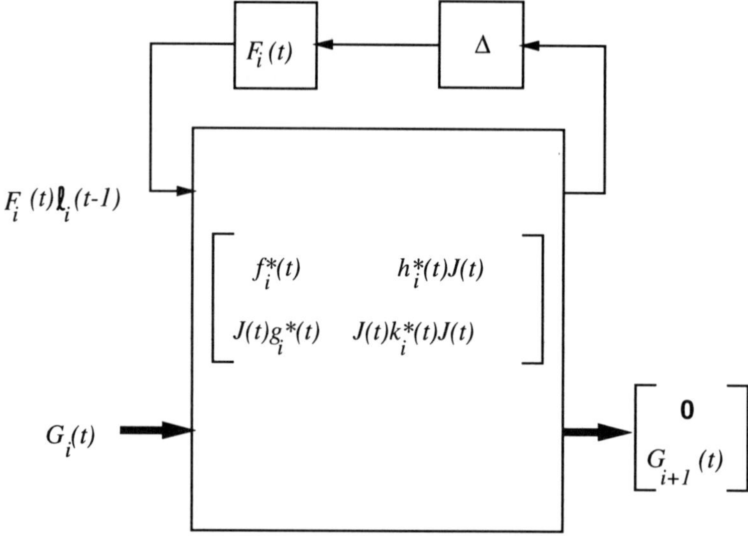

FIG. 5.1. *Time-variant transmission-line structure of the recursive algorithm.*

where $\Theta_i(t)$ is an arbitrary $J(t)$-unitary matrix, and $\tau_i(t)$ is an arbitrary complex number chosen on the circle $|\tau_i(t)|^2 = d_i(t-1)/d_i(t)$.

5.1. Time-variant derivatives. Before discussing the application of the above algorithm to time-variant interpolation and matrix completion problems, we first introduce some notation and extend the notion of "derivatives" to the time-variant setting [20,21]. We consider a finite-dimensional linear *time-variant* state-space model with a bounded upper-triangular transfer operator T. The matrix entries of T are denoted by T_{ij} (of dimensions $r(i) \times r(j)$) and correspond to the time-variant Markov parameters of the underlying state-space model:

$$T = \begin{bmatrix} \ddots & \ddots & & & \\ & T_{-1,-1} & T_{-1,0} & T_{-1,1} & \cdots \\ & & \boxed{T_{00}} & T_{01} & T_{02} & \cdots \\ & \text{O} & & T_{11} & T_{12} & T_{13} \\ & & & & \ddots & \ddots \end{bmatrix},$$

where $\boxed{T_{00}}$ denotes the $(0,0)$ entry of T. We further consider a *stable* sequence of scalar points $\{f(t)\}_{t \in \mathbf{Z}}$ ($\mathbf{Z}$ is the set of integers), viz., $\exists\ c > 0$ such that $|f(t)| < c < 1$ for all t. We also introduce the symmetric functions $s_k^{(n)}$ of n variables (taken k at a time). That is, $s_0^{(n)} = 1$ and

$$s_k^{(n)}(x_1, x_2, \ldots, x_n) = \sum_{1 \le i_1 < \ldots < i_k \le n} x_{i_1} x_{i_2} \ldots x_{i_k}.$$

For a uniformly bounded sequence of $1 \times r(t)$ row vectors $\{u(t)\}_{t \in \mathbf{Z}}$, viz., $\exists\ \bar{c} > 0$ such that $\|u(t)\| < \bar{c}$ for all t, we define the $1 \times r(t)$ row vector $u(t) \bullet \mathcal{T}(f(t))$ as follows

$$u(t) \bullet \mathcal{T}(f(t)) = u(t)T_{tt} + f(t)u(t-1)T_{t-1,t} + f(t)f(t-1)u(t-2)T_{t-2,t} + \ldots .$$

This corresponds to a time-variant tangential evaluation along the direction defined by $u(t)$. More generally, we define the $1 \times r(t)$ row vectors (for $p \geq 0$)

$$u(t) \bullet \tfrac{1}{p!} \mathcal{T}^{(p)}(f(t)) \equiv \sum_{m=0}^{\infty} s_m^{(m+p)}[f(t), f(t-1),$$
$$\ldots, f(t - m - p + 1)]u(t - m - p)T_{t-m-p,t} .$$

We shall also use the compact notation $\begin{bmatrix} u_1(t) & u_2(t) \end{bmatrix} \bullet \mathcal{H}_{\mathcal{T}}^2(f(t))$ to denote the row vector $\begin{bmatrix} u_1(t) \bullet \mathcal{T}(f(t)) & u_1(t) \bullet \tfrac{1}{1!}\mathcal{T}^{(1)}(f(t)) + u_2(t) \bullet \mathcal{T}(f(t)) \end{bmatrix}$, which we also write as

$$\begin{bmatrix} u_1(t) & u_2(t) \end{bmatrix} \bullet \begin{bmatrix} \mathcal{T}(f(t)) & \tfrac{1}{1!}\mathcal{T}^{(1)}(f(t)) \\ & \mathcal{T}(f(t)) \end{bmatrix} .$$

More generally, we write $\begin{bmatrix} u_1(t) & u_2(t) & \ldots & u_r(t) \end{bmatrix} \bullet \mathcal{H}_{\mathcal{T}}^r(f(t)) =$

$$\begin{bmatrix} u_1(t) & u_2(t) & \ldots & u_r(t) \end{bmatrix} \bullet$$

$$\begin{bmatrix} \mathcal{T}(f(t)) & \tfrac{1}{1!}\mathcal{T}^{(1)}(f(t)) & \tfrac{1}{2!}\mathcal{T}^{(2)}(f(t)) & \cdots & \tfrac{1}{(r-1)!}\mathcal{T}^{(r-1)}(f(t)) \\ & \mathcal{T}(f(t)) & \tfrac{1}{1!}\mathcal{T}^{(1)}(f(t)) & \cdots & \tfrac{1}{(r-2)!}\mathcal{T}^{(r-2)}(f(t)) \\ & & \ddots & & \vdots \\ & \mathbf{O} & & \mathcal{T}(f(t)) & \tfrac{1}{1!}\mathcal{T}^{(1)}(f(t)) \\ & & & & \mathcal{T}(f(t)) \end{bmatrix} .$$

We remark that the above expressions for time-variant derivatives and tangential evaluation reduce to the standard definitions in the time-invariant case, where $\mathcal{T}$ is a Toeplitz operator.

6. Time-variant Hermite-Fejér. We now extend the statement of the Hermite-Fejér problem to the time-variant setting. This extension includes as special cases the time-variant versions of the Carathéodory-Fejér and Nevanlinna-Pick problems studied in [16,18,19].

We consider m stable points $\{\alpha_i(t)\}_{i=0}^{m-1}$ inside the open unit disc, and we associate with each point $\alpha_i(t)$ a positive integer $r_i \geq 1$ and uniformly bounded row vectors $\mathbf{a}_i(t)$ and $\mathbf{b}_i(t)$ partitioned as follows

$$\mathbf{a}_i(t) = \begin{bmatrix} u_1^{(i)}(t) & u_2^{(i)}(t) & \ldots & u_{r_i}^{(i)}(t) \end{bmatrix}, \quad \mathbf{b}_i(t) = \begin{bmatrix} v_1^{(i)}(t) & v_2^{(i)}(t) & \ldots & v_{r_i}^{(i)}(t) \end{bmatrix} ,$$

where $u_j^{(i)}(t)$ and $v_j^{(i)}(t)$ $(j = 1, \ldots, r_i)$ are $1 \times p(t)$ and $1 \times q(t)$ row vectors, respectively. The time-variant Hermite-Fejér interpolation problem then reads as follows.

Problem 6.1. Given m stable points $\{\alpha_i(t)\}$ with the associated data $r_i, \mathbf{a}_i(t)$, and $\mathbf{b}_i(t)$, describe all upper triangular strictly contractive transfer operators $\mathcal{S}$ ($\|\mathcal{S}\|_\infty < 1$) that satisfy

$$(6.1) \qquad \mathbf{b}_i(t) = \mathbf{a}_i(t) \bullet \mathcal{H}_{\mathcal{S}}^{r_i}(\alpha_i(t)) \quad \text{for} \quad 0 \leq i \leq m - 1.$$

The first step in the solution consists in constructing three matrices $F(t), G(t)$, and $J(t)$ directly from the interpolation data as in the time-invariant case: we define $J(t) = (I_{p(t)} \oplus -I_{q(t)})$, and associate with each $\alpha_i(t)$ a Jordan block $\bar{F}_i(t)$ of size $r_i \times r_i$,

$$\bar{F}_i(t) = \begin{bmatrix} \alpha_i(t) & & & \\ 1 & \alpha_i(t) & & \\ & \ddots & \ddots & \\ & & 1 & \alpha_i(t) \end{bmatrix},$$

and two $r_i \times p(t)$ and $r_i \times q(t)$ matrices $U_i(t)$ and $V_i(t)$, respectively, which are composed of the row vectors associated with $\alpha_i(t)$,

$$U_i(t) = \begin{bmatrix} u_1^{(i)}(t) \\ u_2^{(i)}(t) \\ \vdots \\ u_{r_i}^{(i)}(t) \end{bmatrix} \quad \text{and} \quad V_i(t) = \begin{bmatrix} v_1^{(i)}(t) \\ v_2^{(i)}(t) \\ \vdots \\ v_{r_i}^{(i)}(t) \end{bmatrix}.$$

Then $F(t) = \text{diagonal} \{\bar{F}_0(t), \bar{F}_1(t), \ldots, \bar{F}_{m-1}(t)\}$ and

$$(6.2) \qquad G(t) = \begin{bmatrix} U_0(t) & V_0(t) \\ U_1(t) & V_1(t) \\ \vdots & \vdots \\ U_{m-1}(t) & V_{m-1}(t) \end{bmatrix} \equiv [\, \mathbf{U}(t)\ \mathbf{V}(t) \,].$$

Let $n = \sum_{i=0}^{m-1} r_i$ and $r(t) = p(t) + q(t)$, then $F(t)$ and $G(t)$ are $n \times n$ and $n \times r(t)$ matrices respectively. We shall denote the diagonal entries of $F(t)$ by $\{f_i(t)\}_{i=0}^{n-1}$ (for example, $f_0(t) = f_1(t) = \ldots = f_{r_0-1}(t) = \alpha_0(t)$). We also associate with the interpolation problem the time-variant displacement equation

$$(6.3) \qquad R(t) - F(t)R(t-1)F^*(t) = G(t)J(t)G^*(t).$$

We shall further assume that the interpolation data satisfy the following nondegeneracy condition, which is automatically satisfied in many problems,

$$\mathcal{U}(t) \equiv [\, \ldots F(t)F(t-1)\mathbf{U}(t-2)\ F(t)\mathbf{U}(t-1)\ \mathbf{U}(t) \,]$$

has the property,

$$(6.4) \qquad \mathcal{U}(t)\mathcal{U}^*(t) > \mu > 0 \quad \text{for all } t \ ,$$

where μ is a fixed constant. The more general case is treated in [21] and will be briefly discussed later. The proof of the next theorem follows the same lines as that of Theorem 4.2 [20].

THEOREM 6.2. *The tangential Hermite-Fejér problem is solvable if, and only if, there exists a real number $\epsilon > 0$, independent of t, such that the solution $R(t)$ of (6.3) satisfies $R(t) > \epsilon I$ for all t.*

Before proceeding further, we first state [20] the implications of stability and uniform boundedness of the interpolation data $\{f_i(t),\ \mathbf{a}_i(t),\ \mathbf{b}_i(t)\}$ on the boundedness of the quantities $d_i(t)$ and $g_i(t)$ that are needed in the recursive procedure.

LEMMA 6.3. *The sequences $\{d_i(t)\}_{t\in\mathbf{Z}}$ and $\{g_i(t)\}_{t\in\mathbf{Z}}$ obtained through the recursive Schur reduction procedure are uniformly bounded. More specifically, there exist real numbers b_d, c_d, and c_g (independent of t) such that*

$$0 < b_d < d_i(t) < c_d \quad \text{and} \quad \|g_i(t)\| < c_g \quad \text{for all } t.$$

6.1. Interpolation properties. Observe again that each generator step as in (5.2) involves a linear first-order discrete-time system (in state-space form) that appears on the right-hand side of (5.2), viz.,

$$(6.5) \quad \begin{bmatrix} \mathbf{x}_i(t+1) & \mathbf{y}_i(t) \end{bmatrix} = \begin{bmatrix} \mathbf{x}_i(t) & \mathbf{w}_i(t) \end{bmatrix} \begin{bmatrix} f_i^*(t) & h_i^*(t)J(t) \\ J(t)g_i^*(t) & J(t)k_i^*(t)J(t) \end{bmatrix} \ ,$$

where $\mathbf{x}_i(t)$ is the state and $\mathbf{w}_i(t)$ is a $1 \times r(t)$ *row* input vector at time t. As in the time-invariant case, the first-order system (6.5) also has important $J(t)$-losslessness and blocking properties.

Let $\mathcal{T}_i = \left[T_{lj}^{(i)} \right]_{l,j=-\infty}^{\infty}$ denote the upper-triangular transfer operator of the ith section (6.5), where $T_{lj}^{(i)}$ denote the $r(l) \times r(j)$ time-variant Markov parameters of $\mathcal{T}_i$ and are given by

$$\begin{aligned}
T_{ll}^{(i)} &= J(l)k_i^*(l)J(l) \ , \\
T_{l,l+1}^{(i)} &= J(l)g_i^*(l)h_i^*(l+1)J(l+1) \ , \\
T_{lj}^{(i)} &= J(l)g_i^*(l)f_i^*(l+1)f_i^*(l+2)\ldots f_i^*(j-1)h_i^*(j)J(j) \ , \\
&\qquad \text{for } j > l+1 \ .
\end{aligned}$$

After n recursive steps (recall that $G(t)$ has n rows) we obtain a cascade of sections $\mathcal{T}$ defined by (Figure 6.1)

$$(6.6) \qquad \mathcal{T} = \mathcal{T}_0\mathcal{T}_1\ldots\mathcal{T}_{n-1}.$$

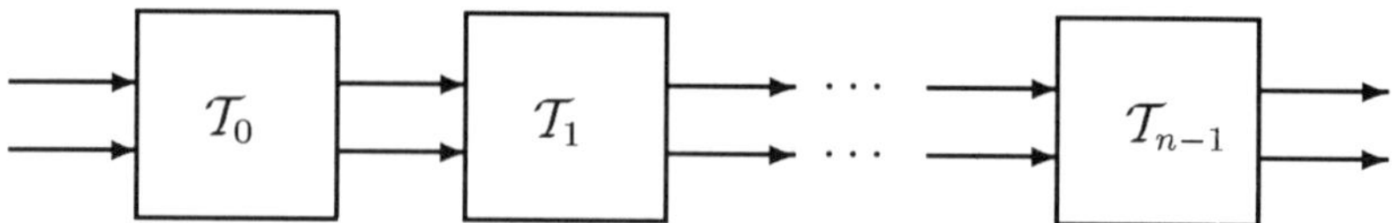

FIG. 6.1. *Cascade of first-order time-variant sections.*

Our purpose is to prove that all solutions $\mathcal{S}$ to the interpolation Problem 6 can be parametrized in terms of a linear fractional transformation based on $\mathcal{T}$.

We already know that $\{f_i(t), g_i(t)\}_{t \in \mathbf{Z}}$ are stable and uniformly bounded sequences (recall Lemma 6.3). Moreover, it is always possible [20,21] to choose the free parameters $\Theta_i(t)$ and $\tau_i(t)$ in (5.5) so as to guarantee the uniform boundedness of the sequences $\{h_i(t), k_i(t)\}_{t \in \mathbf{Z}}$. This is, for example, the case if we set $\Theta_i(t) = I_r$ and choose $\tau_i(t)$ on the circle of radius $\sqrt{\frac{d_i(t-1)}{d_i(t)}}$ but in the opposite direction of $f_i(t)$. Other choices are also possible and lead to time-variant lattice structures [20].

It is then a standard result that the boundedness of $\{f_i(t), g_i(t), h_i(t), k_i(t)\}$ assures the boundedness of the corresponding operator $\mathcal{T}_i$ (see, e.g., [41]). Moreover, if we define the direct sum $\mathcal{J} = \underset{t \in \mathbf{Z}}{\oplus} J(t)$, then it readily follows from the embedding relation (5.3) that each $\mathcal{T}_i$ also satisfies the following $\mathcal{J}$-losslessness property,

$$\mathcal{T}_i \mathcal{J} \mathcal{T}_i^* = \mathcal{J} \quad \text{and} \quad \mathcal{T}_i^* \mathcal{J} \mathcal{T}_i = \mathcal{J}$$

Furthermore, each section $\mathcal{T}_i$ satisfies an important time-variant blocking property.

THEOREM 6.4. *Each first-order section $\mathcal{T}_i$ satisfies*

$$\left[\ldots f_i(t)f_i(t-1)g_i(t-2) \; f_i(t)g_i(t-1) \; g_i(t) \; ? \right] \mathcal{T}_i = \left[\mathbf{0} \; ? \right] \; ,$$

where $g_i(t)$ is at the tth position of the row vector. Consequently, $g_i(t) \bullet \mathcal{T}_i(f_i(t)) = \mathbf{0}$.

Proof. This follows directly from the embedding result (5.3) (as well as from the fact that each step of the generator recursion (5.2) produces a zero row). The output of $\mathcal{T}_i$ at time t is given by

$$\mathbf{y}_i(t) = \ldots + f_i(t)f_i(t-1)g_i(t-2)T_{t-2,t} + f_i(t)g_i(t-1)T_{t-1,t} + g_i(t)T_{tt}$$
$$= [-d_i(t-1) + d_i(t-1)] \, f_i(t)h_i^*(t)J(t) = \mathbf{0} \; ,$$

where we substituted the expressions for the Markov parameters and used

$$d_i(t) = g_i(t)J(t)g_i^*(t) + f_i(t)g_i(t-1)J(t-1)g_i^*(t-1)f_i^*(t)+$$
$$f_i(t)f_i(t-1)g_i(t-2)J(t-2)g_i^*(t-2)f_i^*(t-1)f_i^*(t) + \dots$$

The same argument holds for the previous outputs. $\quad\square$

The $\mathcal{J}$-losslessness and blocking properties of each section T_i reflect on the entire cascade T, and it readily follows that T is a bounded upper-triangular linear operator that satisfies $T\mathcal{J}T^* = T^*\mathcal{J}T = \mathcal{J}$. It also follows from the last theorem that T satisfies an important global blocking property.

THEOREM 6.5. *The entire cascade T satisfies the global blocking property*

$$(6.7) \quad \left[\dots F(t)F(t-1)G(t-2)\ F(t)G(t-1)\ G(t)\ \mathbf{0}\ \mathbf{0} \dots\right]T = \left[\mathbf{0}\ ?\right],$$

where $G(t)$ is in the tth position. That is, if we apply to T the block input

$$\tilde{\mathbf{U}}(t) = \left[\dots F(t)F(t-1)G(t-2)\ F(t)G(t-1)\ G(t)\ \mathbf{0}\ \mathbf{0} \dots\right]$$

then the output is zero up to and including time t.

Proof. This follows from the generator recursion (5.2) and from the Jordan structure of $F(t)$. When the first row of $\tilde{\mathbf{U}}(t)$ goes through the first section T_0, it annihilates the output of the entire cascade T due to the blocking property of T_0. When the second row of $\tilde{\mathbf{U}}(t)$ goes through T_0, we obtain at the output of T_0 (as a consequence of (5.2) and the Jordan structure of $F(t)$) a zero-direction vector for T_1, which again annihilates the output of the entire cascade T, and so on. $\quad\square$

Expression (6.7) is closely related to the interpolation conditions of Problem 6. To motivate this, we denote by $s_i = \sum_{p=0}^{i-1} r_p$, $s_0 = 0$, the total size of the Jordan blocks prior to $\bar{F}_i(t)$. By comparing terms on both sides of (6.7) (and by using the Jordan structure of $F(t)$) we can verify that (6.7) can be rewritten in the following form

$$(6.8) \quad \left[e_{s_i}G(t)\ e_{s_i+1}G(t) \dots e_{s_i+r_i-1}G(t)\right] \bullet \mathcal{H}_T^{r_i}(\alpha_i(t)) = 0,$$

where the row vector on the left hand-side of (6.8) is composed of the r_i row vectors in $\left[U_i(t)\ V_i(t)\right]$ associated with $\alpha_i(t)$, viz.,

$$\left[u_1^{(i)}(t)\ v_1^{(i)}(t)\ u_2^{(i)}(t)\ v_2^{(i)}(t) \dots u_{r_i}^{(i)}(t)\ v_{r_i}^{(i)}(t)\right].$$

We now show how to parametrize all solutions to the interpolation problem in terms of T. If we partition the matrix entries T_{lj} of the cascade

T accordingly with $J(l)$ and $J(j)$,

$$T_{lj} = \begin{bmatrix} T_{11}^{lj} & T_{12}^{lj} \\ T_{21}^{lj} & T_{22}^{lj} \end{bmatrix} \, ,$$

and consider the triangular operators

$$\mathcal{T}_{12} = \left[T_{12}^{lj} \right]_{l,j=-\infty}^{\infty} \quad \text{and} \quad \mathcal{T}_{22}^{(i)} = \left[T_{22}^{lj} \right]_{l,j=-\infty}^{\infty} .$$

Then it can be shown [20,21] that $\mathcal{S} = -\mathcal{T}_{12}\mathcal{T}_{22}^{-1}$ is an upper-triangular strictly contractive operator. It also follows from Theorem 6.5 that $\mathcal{S}$ satisfies the required interpolation conditions. For instance, we conclude from (6.8) that

$$\left[\ldots f_0(t)f_0(t-1)g_0(t-2) \; f_0(t)g_0(t-1) \; g_0(t) \; \mathbf{0} \; \mathbf{0} \ldots \right] \begin{bmatrix} \mathcal{T}_{12} \\ \mathcal{T}_{22} \end{bmatrix} = \begin{bmatrix} \mathbf{0} & ? \end{bmatrix} \, ,$$

or equivalently,

$$v_1^{(0)}(t) = u_1^{(0)}(t) \bullet \mathcal{S}(f_0(t)).$$

This argument can be continued, as in the time-invariant case, to show that $\mathcal{S}$ satisfies the remaining interpolation conditions. Moreover, we can parametrize all solutions as follows [20,21].

THEOREM 6.6. *All solutions $\mathcal{S}$ to the tangential Hermite-Fejér problem are given through a linear fractional transformation of a strictly contractive upper-triangular operator $\mathcal{K}$,*

$$(6.9) \qquad \mathcal{S} = - \left[\mathcal{T}_{11}\mathcal{K} + \mathcal{T}_{12} \right] \left[\mathcal{T}_{21}\mathcal{K} + \mathcal{T}_{22} \right]^{-1} .$$

7. Completion problems. The discussion in the previous sections was restricted to structured matrices R with scalar entries r_{mj}. The results however, are more general. We now state a result proved in [21], which shows that, under a certain positivity condition, it is always possible to associate an abstract interpolation problem with a time-variant structured matrix, and that the existence of a solution to this abstract problem is characterized by a natural embedding property. This result includes as particular cases earlier developments in [9,42,43].

We consider a family of block matrices, depending on the parameter $t \in \mathbf{Z}$,

$$R(t) = [r_{mj}(t)]_{m,j=0}^{n-1} \, , \quad r_{mj}(t) \in \mathcal{L}(\mathcal{H}_j(t), \mathcal{H}_m(t)) \, ,$$

where $\{\mathcal{H}_m(t)\}_{m=0}^{n-1}$ are families of Hilbert spaces depending on the parameter $t \in \mathbf{Z}$, and for two Hilbert spaces $\mathcal{H}$ and $\mathcal{H}'$ we use the notation

$\mathcal{L}(\mathcal{H}, \mathcal{H}')$ to denote the set of linear bounded operators acting from $\mathcal{H}$ into $\mathcal{H}'$. If we define $\mathcal{H}(t) = \overset{n-1}{\underset{m=0}{\oplus}} \mathcal{H}_m(t)$, then $R(t) \in \mathcal{L}(\mathcal{H}(t))$. We also consider two families $\{\mathcal{F}(t)\}_{t\in\mathbf{Z}}$ and $\{\mathcal{G}(t)\}_{t\in\mathbf{Z}}$ of Hilbert spaces, two families of bounded linear operators,

$$F(t) \in \mathcal{L}(\mathcal{H}(t-1), \mathcal{H}(t)) \quad \text{and} \quad G(t) \in \mathcal{L}(\mathcal{F}(t) \oplus \mathcal{G}(t), \mathcal{H}(t)),$$

and we define the symmetry $J(t) = (I_{\mathcal{F}(t)} \oplus -I_{\mathcal{G}(t)})$, on $\mathcal{F}(t) \bigoplus \mathcal{G}(t)$, where $I_{\mathcal{F}(t)}$ denotes the identity operator on the space $\mathcal{F}(t)$. We also write $G(t) = \left[\, \mathbf{U}(t)\ \mathbf{V}(t)\,\right]$, where

$$\mathbf{U}(t) \in \mathcal{L}(\mathcal{F}(t), \mathcal{H}(t)) \quad \text{and} \quad \mathbf{V}(t) \in \mathcal{L}(\mathcal{G}(t), \mathcal{H}(t)).$$

We assume that $\{F(t)\}_{t\in\mathbf{Z}}$ is a uniformly bounded family of lower triangular operators, with stable diagonal entries $\{f_0(t), f_1(t), \ldots, f_{n-1}(t)\}$, viz.,

$$\exists\ c_f > 0 \quad \text{such that}\ \|f_i(t)\| \le c_f < 1$$
$$\text{for all}\ t \in \mathbf{Z}\ \text{and}\ i = 0, 1, \ldots, n-1.$$

We shall say that $\{F(t)\}_{t\in\mathbf{Z}}$ is a *stable* family (these conditions can be relaxed [21]). We also assume that $\{G(t)\}_{t\in\mathbf{Z}}$ is a uniformly bounded family of operators, viz.,

$$\exists\ c_G > 0 \quad \text{such that}\ \|G(t)\| \le c_G \quad \text{for all}\ t \in \mathbf{Z}.$$

We shall also say that $\{R(t)\}_{t\in\mathbf{Z}}$ has a time-variant displacement structure with respect to the family of operators $\{F(t), G(t), J(t)\}_{t\in\mathbf{Z}}$ if $\{R(t)\}_{t\in\mathbf{Z}}$ satisfies the time-variant displacement equation

$$(7.1) \qquad R(t) - F(t)R(t-1)F^*(t) = G(t)J(t)G^*(t) \ ,$$

where the symbol $*$ refers to the adjoint operator $(F^*(t) = F(t)^*)$. The cardinal number, $r(t) = \dim \mathcal{F}(t) + \dim \mathcal{G}(t)$, is called the displacement rank of $R(t)$ in (7.1). We say that (7.1) has a *Pick solution* if, and only if, $R(t)$ is positive-semidefinite for every $t \in \mathbf{Z}$. If we define,

$$\mathcal{U}(t) = \left[\, \ldots F(t)F(t-1)\mathbf{U}(t-2)\ F(t)\mathbf{U}(t-1)\ \mathbf{U}(t)\,\right]$$

and

$$\mathcal{V}(t) = \left[\, \ldots F(t)F(t-1)\mathbf{V}(t-2)\ F(t)\mathbf{V}(t-1)\ \mathbf{V}(t)\,\right] \ ,$$

then we remark that $\mathcal{U}(t)$ and $\mathcal{V}(t)$ are well defined bounded linear operators,

$$\mathcal{U}(t) \in \mathcal{L}(\underset{j\le t}{\oplus} \mathcal{F}(j), \mathcal{H}(t)), \quad \mathcal{V}(t) \in \mathcal{L}(\underset{j\le t}{\oplus} \mathcal{G}(j), \mathcal{H}(t)) \ ,$$

and that $R(t)$ is given by

$$(7.2) \qquad R(t) = \mathcal{U}(t)\mathcal{U}^*(t) - \mathcal{V}(t)\mathcal{V}^*(t).$$

The following result [21] shows that the existence of a Pick solution of (7.1) is equivalent to the existence of an upper triangular contraction relating $\mathcal{U}(t)$ and $\mathcal{V}(t)$. We have already encountered a special case of this result in the proof of Theorem 4.2 (and also Theorem 6.2).

THEOREM 7.1. *The time-variant displacement equation (7.1) has a Pick solution $R(t)$ if, and only if, there exists an upper triangular contraction $\mathcal{S}$ ($\|\mathcal{S}\| \le 1$),*

$$\mathcal{S} \in \mathcal{L}(\bigoplus_{t \in \mathbf{Z}} \mathcal{G}(t), \ \bigoplus_{t \in \mathbf{Z}} \mathcal{F}(t)),$$

such that

$$(7.3) \qquad \mathcal{V}(t) = \mathcal{U}(t) P_{\mathcal{F}}(t) \mathcal{S} / \bigoplus_{j \le t} \mathcal{G}(j) \quad \text{for every } t \in \mathbf{Z}$$

where $P_{\mathcal{F}}(t)$ denotes the orthogonal projection of $\bigoplus_{t \in \mathbf{Z}} \mathcal{F}(t)$ onto $\bigoplus_{j \le t} \mathcal{F}(j)$.

It can be further shown that the contraction $\mathcal{S}$ is the solution of a general interpolation problem [21]. We shall instead show that several completion problems recently considered in connection with moment theory can be solved within the framework of displacement structure theory.

We fix a family $\{\mathcal{E}(n)\}_{n \in \mathbf{Z}}$ of Hilbert spaces and a positive integer p. We now verify that the solution of the following band completion problem [44] follows as a special case of Theorem 7.1.

Problem 7.2. Given a family $\{\tilde{Q}_{ij} / \ i, j \in \mathbf{Z}, \ |j - i| \le p\}$ of operators such that $\tilde{Q}_{ij} = \tilde{Q}_{ji}^*$ and $\tilde{Q}_{ij} \in \mathcal{L}(\mathcal{E}(j), \mathcal{E}(i))$, it is required to find conditions for the existence of a positive definite kernel $M = [Q_{ij}]_{i,j \in \mathbf{Z}}$ such that for $i, j \in \mathbf{Z}$ and $|j - i| \le p$, $Q_{ij} = \tilde{Q}_{ij}$.

By a positive definite kernel we mean an application $M = [Q_{ij}]_{i,j \in \mathbf{Z}}$ on $\mathbf{Z} \times \mathbf{Z}$ such that for $i, j \in \mathbf{Z}$, we have

$$Q_{ij} \in \mathcal{L}(\mathcal{E}(j), \mathcal{E}(i)) \quad \text{and} \quad \sum_{i,j=-n}^{n} < Q_{ij} h_j, h_i > \ \ge 0 \ ,$$

for every integer $n > 0$ and every set of vectors $\{h_{-n}, h_{-n+1}, \ldots, h_n\}$, $h_k \in \mathcal{E}(k)$, $|k| \le n$. In case $\tilde{Q}_{ij} = \tilde{Q}_{|j-i|}$, the above problem is the well-known truncated trigonometric moment problem. The case $\mathcal{E}(n) = \mathbf{0}$ for $|n|$ large enough, was solved in [44].

Without loss of generality, we can suppose $\tilde{Q}_{ii} = I$ for all $i \in \mathbf{Z}$. Define the spaces,

$$(7.4) \qquad \mathcal{H}(t) = \bigoplus_{k=0}^{p} \mathcal{E}(-t+k), \quad \mathcal{F}(t) = \mathcal{G}(t) = \mathcal{E}(-t) \ ,$$

and the operators

$$(7.5) \qquad \mathbf{U}(t) = \begin{bmatrix} I \\ \tilde{Q}_{-t+1,-t} \\ \tilde{Q}_{-t+2,-t} \\ \vdots \\ \tilde{Q}_{-t+p,-t} \end{bmatrix} \quad \text{and} \quad \mathbf{V}(t) = \begin{bmatrix} 0 \\ \tilde{Q}_{-t+1,-t} \\ \tilde{Q}_{-t+2,-t} \\ \vdots \\ \tilde{Q}_{-t+p,-t} \end{bmatrix}.$$

We also consider the operators $J(t) = (I_{\mathcal{F}(t)} \oplus -I_{\mathcal{G}(t)})$,

$$(7.6) \qquad F(t) = \begin{bmatrix} 0 \\ I & 0 \\ & \ddots & \ddots \\ & & I & 0 \end{bmatrix}, \quad G(t) = \begin{bmatrix} \mathbf{U}(t) & \mathbf{V}(t) \end{bmatrix}.$$

These elements specify a displacement structure of the form (7.1), and the following result follows from Theorem 7.1.

THEOREM 7.3. *Problem 7 has solutions if, and only if, the displacement equation associated with the data (7.4)–(7.6), has a Pick solution.*

The following so-called tangential Carathéodory-Fejér) problem is a special case of Problem 6 and arises, for example, in model validation [45].

Problem 7.4. Given families of matrices $\{U_i(t), V_i(t)\}_{t \in \mathbf{Z}}$, $i = 0, 1,$ $\ldots, n - 1$, it is required to find conditions for the existence of an upper triangular contraction $\mathcal{S}$ such that

$$\begin{bmatrix} U_0(t-n+1) \ldots U_{n-1}(t) \end{bmatrix} \begin{bmatrix} \mathcal{S}_{t-n+1,t-n+1} & & \cdots & \mathcal{S}_{t-n+1,t} \\ & \ddots & & \vdots \\ & & \mathcal{S}_{t-1,t-1} & \mathcal{S}_{t-1,t} \\ & O & & \mathcal{S}_{tt} \end{bmatrix} =$$

$$\begin{bmatrix} V_0(t-n+1) \ldots V_{n-1}(t) \end{bmatrix}.$$

This problem can be stated as imposing linear constraints on the "time-variant derivatives" of $\mathcal{S}$. To reduce the problem to our framework we define for all t,

$$\mathbf{U}(t) = \begin{bmatrix} U_0(t) \\ U_1(t) \\ \vdots \\ U_{n-1}(t) \end{bmatrix}, \quad \mathbf{V}(t) = \begin{bmatrix} V_0(t) \\ V_1(t) \\ \vdots \\ V_{n-1}(t) \end{bmatrix},$$

$$F(t) = \begin{bmatrix} 0 \\ I & 0 \\ & \ddots & \ddots \\ & & I & 0 \end{bmatrix}, \quad G(t) = \begin{bmatrix} \mathbf{U}(t) & \mathbf{V}(t) \end{bmatrix}.$$

THEOREM 7.5. *The tangential Carathéodory-Fejér problem has solutions if, and only if, the displacement equation associated with the above data has a Pick solution. This is equivalent to the condition*

$$\mathcal{U}(t)\mathcal{U}^*(t) \geq \mathcal{V}(t)\mathcal{V}^*(t) \quad \text{for all } t \in \mathbf{Z} \,,$$

with

$$\mathcal{U}(t) = \begin{bmatrix} & & & U_0(t) \\ & & U_0(t-1) & U_1(t) \\ & & & \vdots \\ U_0(t-n+1) & \dots & U_{n-2}(t-1) & U_{n-1}(t) \end{bmatrix};$$

$$\mathcal{V}(t) = \begin{bmatrix} & & & V_0(t) \\ & & V_0(t-1) & V_1(t) \\ & & & \vdots \\ V_0(t-n+1) & \dots & V_{n-2}(t-1) & V_{n-1}(t) \end{bmatrix}.$$

Another completion problem whose solution can be obtained as a special case of Theorem 7.1 is the so-called strong Parrott problem [46,47].

Problem 7.6. Given matrices B_{ij}, $1 \leq j \leq i \leq n$, $S = \begin{bmatrix} S_1 & S_2 & \dots & S_n \end{bmatrix}$ and $T = \begin{bmatrix} T_1 & T_2 & \dots & T_n \end{bmatrix}$, it is required to find conditions for the existence of a contraction $\mathcal{T}$ of the form

$$\mathcal{T} = \begin{bmatrix} B_{11} & & & \\ B_{21} & B_{22} & & \Large? \\ \vdots & & \ddots & \\ B_{n1} & B_{n2} & \dots & B_{nn} \end{bmatrix},$$

such that $S\mathcal{T} = T$, where ? denotes unspecified entries.

To put this problem into our framework, we define

$$\mathbf{U}(t) = \begin{bmatrix} 0 \\ I \\ 0 \\ \vdots \\ 0 \end{bmatrix}, \quad 1 \leq t \leq n-1, \quad \mathbf{U}(t) = \begin{bmatrix} S_{n+t} \\ I \\ 0 \\ \vdots \\ 0 \end{bmatrix}, \quad -n+1 \leq t \leq 0 \,,$$

$$\mathbf{V}(0) = \begin{bmatrix} T_1 \\ B_{n1} \\ B_{n-1,1} \\ \vdots \\ B_{11} \end{bmatrix}, \quad \mathbf{V}(1) = \begin{bmatrix} T_2 \\ \mathbf{0} \\ B_{n2} \\ \vdots \\ B_{22} \end{bmatrix}, \quad \mathbf{V}(2) = \begin{bmatrix} T_3 \\ \mathbf{0} \\ \mathbf{0} \\ B_{n3} \\ \vdots \\ B_{33} \end{bmatrix}, \ \ldots,$$

$$\mathbf{V}(n-1) = \begin{bmatrix} T_n \\ \mathbf{0} \\ \vdots \\ \mathbf{0} \\ B_{nn} \end{bmatrix},$$

$$F(t) = \begin{bmatrix} I \\ \mathbf{0}\ \mathbf{0} \\ I\ \mathbf{0} \\ I\ \ \mathbf{0} \\ \ddots\ \ddots \\ I\ \ \mathbf{0} \end{bmatrix}, \quad G(t) = \begin{bmatrix} \mathbf{U}(t)\ \mathbf{V}(t) \end{bmatrix}, \quad \text{for } -n+1 \leq t \leq n-1 \ ,$$

and all the elements equal to zero for the other time indices. We then have the following result (using the relaxed version of Theorem 7.1 in [21]).

THEOREM 7.7. *The strong Parrott problem has solutions if, and only if, the time-variant displacement equation associated with the above data has a Pick solution.*

8. Concluding remarks. We discussed several applications of the displacement structure concept to interpolation and matrix completion problems. We emphasized that a transmission-line cascade arises naturally in the study of fast factorization algorithms for structured matrices, and that it has physically meaningful blocking properties or transmission zeros. This simple fact was then exploited to solve interpolation problems in both the time-variant and time-invariant settings. We also showed how several completion or moment problems fit naturally into the framework discussed in this paper.

Our recursive interpolation solution constructs a time-variant transmission-line cascade by *implicitly* considering the triangular factorization of the Pick matrix $R(t)$; unlike several earlier solutions, we do not require explicit knowledge of the matrices $R(t)$ or $R^{-1}(t)$. The whole recursive procedure works only with the matrices $F(t)$ and $G(t)$ that are constructed directly from the interpolation data. The overall computational complexity of the procedure is $O(r(t)n^2)$ operations (additions and multiplications) per time step, where $r(t)$ is a so-called displacement rank (the number of columns of the matrix $G(t)$).

REFERENCES

[1] D.C.YOULA M.SAITO, *Interpolation with positive real functions*, J. Franklin Institute **284** (1967), 77–108.

[2] J.W.HELTON, *Operator theory, analytic functions, matrices and electrical engineering*, CBMS, American Mathematical Society, Providence, RI 1987.

[3] D.SARASON, *Generalized interpolation in* H^∞, American Mathematical Society Transactions **127** (1967), 179–203.

[4] V.M.ADAMJAN, D.Z.AROV, M.G.KREIN, *Infinite hankel and generalized carathéodory-fejér and I. schur problems*, Funktsional Anal. Prilozhen **2** (1968), 1–17. *(Russian)*.

[5] C.FOIAS, A.E.FRAZHO, *The commutant lifting approach to interpolation problems*, Operator Theory: Advances and Applications, Birkhäuser, Basel 1990.

[6] P.I.FEDČINA, *A criterion for the solvability of the nevanlinna-pick tangent problem*, Mat. Issled **7** (1972), 213–227.

[7] PH.DELSARTE, Y.GENIN, Y.KAMP, *The Nevanlinna-Pick problem for matrix-valued functions*, SIAM J. Appl. Math. **36** (1979), 47–61.

[8] J.A.BALL, J.W.HELTON, *A beurling-lax theorem for the lie group* $u(m, n)$ *which contains most classical interpolation theory*, J. Operator Theory **9** (1983), 107–142.

[9] D.ALPAY, H.DYM, *On applications of reproducing kernel spaces to the schur algorithm and rational J-unitary factorization*, Operator Theory: Advances and Applications, (ed, I.Gohberg) **18** 1986, 89–159.

[10] P.DEWILDE, H.DYM, *Lossless chain scattering matrices and optimum linear prediction: the vector case*, Circuit Theory and Applications **9** (1981), 135–175.

[11] H.DYM, *J-contractive matrix functions, reproducing kernel hilbert spaces and interpolation*, CBMS, American Mathematical Society, Providence, RI **71** 1989.

[12] H.KIMURA, *Directional interpolation approach to* H^∞*-optimization and robust stabilization*, IEEE Transactions on Automatic Control **32** (1987), 1085–1093.

[13] J.A.BALL, I.GOHBERG, L.RODMAN, *Interpolation of rational matrix functions*, Operator Theory: Advances and Applications, Birkhäuser, Basel 1990.

[14] D.J.N.LIMEBEER, B.D.O.ANDERSON, *An interpolation theory approach to* H^∞*-controller degree bounds*, Linear Algebra and Its Applications **98** (1988), 347–386.

[15] D.J.N.LIMEBEER, M.GREEN, *Parametric interpolation,* H^∞*-control and model reduction*, Int. J. Control **52** (1990), 293–318.

[16] J.A.BALL, I.GOHBERG, *A commutant lifting theorem for triangular matrices with diverse applications*, Integral Equations and Operator Theory **8** (1985), 205–267.

[17] T.CONSTANTINESCU, *Schur analysis of positive block-matrices*, Operator Theory: Advances and Applications, (ed, I. Gohberg), Birkhäuser, Boston **18** 1986, 191–206.

[18] J.A.BALL, I.GOHBERG, M.A.KAASHOEK, *Nevanlinna-pick interpolation for time-varying input-output maps: the discrete case*, Operator Theory: Advances and Applications, (ed, I. Gohberg), Birkhäuser Verlag Basel **56** 1992, 1–51.

[19] P.DEWILDE, H.DYM, *Interpolation for upper triangular operators*, Operator Theory: Advances and Applications, (ed, I. Gohberg), Birkhäuser Verlag Basel **56** 1992, 153–260.

[20] A.H.SAYED, T.CONSTANTINESCU, T.KAILATH, *Time-variant displacement structure and interpolation problems*, IEEE Transactions on Automatic Control *(to appear)*.

[21] T.CONSTANTINESCU, A.H.SAYED, T.KAILATH, *Displacement structure and some general interpolation problems*, SIAM J. Matrix Anal. and Applications *(to appear)*.

[22] T.KAILATH, J.CHUN, *Generalized displacement structure for block-toeplitz, toeplitz-block, and toeplitz-derived matrices*, SIAM J. Matrix Anal. Appl. **15**

(1) January (1994), 114–128.

[23] H.LEV-ARI, T.KAILATH, *Triangular factorization of structured hermitian matrices*, Operator Theory: Advances and Applications, (ed, I. Gohberg), Birkhäuser, Boston **18** 1986, 301–324.

[24] A.H.SAYED, T.KAILATH, *Fast algorithms for generalized displacement structures and lossless systems (to appear)* Linear Algebra and Its Applications.

[25] M.MORF, T.KAILATH, *Square root algorithms for least squares estimation*, IEEE Transactions on Automatic Control **20** (1975), 487–497.

[26] G.H.GOLUB, C.F.VAN LOAN, *Matrix computations*, The Johns Hopkins University Press, Baltimore, (*second edition*) 1989.

[27] C.M.RADER, A.O.STEINHARDT, *Hyperbolic householder transformations*, IEEE Transactions on Acoustics, Speech and Signal Processing **34** (1986), 1589–1602.

[28] G.CYBENKO, M.BERRY, *Hyperbolic householder algorithms for factoring structured matrices*, SIAM J. Matrix Anal. Appl. **11** (1990), 499–520.

[29] H.LEV-ARI, T.KAILATH, *State-space approach to factorization of lossless transfer functions and structured matrices*, Linear Algebra and Its Applications **162–164** (1992), 273–295.

[30] A.H.SAYED, T.KAILATH, H.LEV-ARI, T.CONSTANTINESCU, *Recursive solutions of rational interpolation problems via fast matrix factorization, (to appear in)* Integral Equations and Operator Theory.

[31] A.H.SAYED, *Displacement structure in signal processing and mathematics*, PhD thesis, Stanford University, Stanford, CA, August 1992.

[32] H.LEV-ARI, *Nonstationary lattice-filter modeling*, PhD thesis, Stanford University, Stanford, CA, December 1983.

[33] N.I.AKHIEZER, *The classical moment problem and some related questions in analysis*, Hafner Publishing Company, NY 1965.

[34] I.SCHUR, *Über potenzreihen die im inneren des einheitskreises beschränkt sind*, Journal für die Reine und Angewandte Mathematik, **147** (1917), 205–232. (*English translation in Operator theory: advances and applications*, **18** (1986), 31–88.)

[35] C.CARATHÉODORY, L.FEJÉR, *Über den zusammenhang der extremen von harmonischen funktionen mit ihren koeffizienten ünd uber den picard-landauschen satz*, Rend. Circ. Mat. Palermo **32** (1911), 218–239.

[36] R.NEVANLINNA, *Über beschränkte funktionen die in gegebenen punkten vorgeschreibene funktionswerte bewirkt werden*, Anal. Acad. Sci. Fenn. **13** (1919), 1–71.

[37] G.PICK, *Über die beschränkungen analytischer funktionen, welche durch vorgegebene funktionswerte bewirkt werden*, Math. Ann. **77** (1916), 7–23.

[38] H.DYM, *On reproducing kernel spaces, J-unitary matrix functions, interpolation and displacement rank*, Operator Theory: Advances and Applications, (ed, I.Gohberg) **41** 1989, 173–239.

[39] D.ALPAY, P.BRUINSMA, A.DIJKSMA, H.DE SNOO, *Interpolation problems, extensions of symmetric operators and reproducing kernel spaces I*, Operator Theory: Advances and Applications **50** (1991), 35–82.

[40] A.H.SAYED, H.LEV-ARI, T.KAILATH, *Time-variant displacement structure and triangular arrays*, IEEE Transactions on Signal Processing, (*to appear*).

[41] I.GOHBERG, M.A.KAASHOEK, *Time-varying linear systems with boundary conditions and integral operators I. the transfer operator and its applications*, Integral Equations and Operator Theory **7** (1984), 325–391.

[42] A.A.NUDELMAN, *On a generalization of classical interpolation problems*, Dokl. Akad. Nauk SSR **256** (1981), 790–793. (*also* Soviet Math. Dokl. **23** (1981), 125–128.)

[43] M.ROSENBLUM, J.ROVNYAK, *Hardy classes and operator theory*, Oxford Univ. Press 1985.

[44] H.DYM, I.GOHBERG, *Extensions of band matrices with band inverses*, Linear Al-

gebra and Its Applications **36** (1981), 1–24.

[45] K.POOLLA, P.KHARGONEKAR, A.TIKKU, J.KRAUSE, K.NAGPAL, *A time-domain approach to model validation*, Proc. American Control Conference, Chicago, IL, **1** (1992), 313–317.

[46] C.FOIAS, A.TANNENBAUM, *A strong parrott theorem*, Proc. of the AMS **106** (1989), 777–784.

[47] M.BAKONYI, H.J.WOERDEMAN, *On the strong parrott completion problem*, Proc. of the AMS (*to appear*).